COMMON SENSE
HEALTHCARE
POLICY FOR
COMMON SENSE
AMERICANS
(AND PRESIDENTIAL CANDIDATES)

BY PETER J. PITTS
PRESIDENT, CENTER FOR MEDICINE IN THE PUBLIC INTEREST

COGNITO
PRESS

First Printing: 2019

ISBN 978-0-578-22444-2

Cognito Press

757 Third Avenue
New York, NY 10017

www.cmpi.org

10/19

To Kerrie Brady —

La Reine de la douleur.

AUTHOR'S NOTE

The articles in this book have previously appeared in a wide variety of newspapers, magazines, blogs, and policy journals. They remain relevant today. Common sense is like that. Good ideas are complicated. They take time to germinate and take root. Acceptance for change has to be earned through honest and clear explanation and regular exhortation. Bad ideas are hard to kill, often because they sound good, lend themselves to soundbites, and everyone (especially politicians!) hope for quick fixes. We have to stop dividing the healthcare debate—especially when it comes to medicines—into good guys and bad guys. There is an apt Japanese proverb, "Don't fix the blame, fix the problem." We're all in this together, and together is how we will succeed in advancing the quality and accessibility of healthcare in the United States.

I would like to dedicate this book to my wife and best friend, Sweet Jane, and also to the dedicated men and women of the U.S. Food and Drug Administration. Never have so many owed so much to so few.

Peter J. Pitts
New York City

CONTENTS

4. PRICE CONTROLS

INTRODUCTION: "THE TRUTH IS RARELY PURE AND NEVER SIMPLE."

5. FIXING THE SYSTEM

INTRODUCTION: "IT'S TIME TO INJECT A LARGE DOSE OF TRANSPARENCY INTO THE SYSTEM."

6. HELPING PATIENTS

INTRODUCTION: "NOTHING IN THE WORLD IS MORE DANGEROUS THAN SINCERE IGNORANCE AND CONSCIENTIOUS STUPIDITY."

INTRODUCTION

"Anyone who isn't confused really doesn't understand the situation." –Edward R. Murrow

Imagine American healthcare spending as a dollar bill divided into 100 pennies. How many pennies do you think represent spending on prescription drugs? Sixty? Eighty? Wrong. The answer is 11.5 (with just under 9 percent being spent on innovative, on-patent medicines). The other 88.5 represent everything else—from doctor visits and hospitalization to administrative charges and insurance.

Put another way, which is the bargain: a hospital stay at about $7,500 a day, or innovative medicines that help keep you healthy and productive? Clearly, fewer cents make more sense.

Yet these and many other facts backing pharmaceuticals as a sound healthcare investment have been twisted to suit the agendas of politicians, pundits, and other competing stakeholders. It goes relatively unreported that insurance companies continue to increase their monthly premiums without really explaining why. The industry claims its costs are increasing because prescription drug costs are busting their budgets. But prescription drugs account for only a small part of monthly insurance-premium hikes. From 1998 to 2003, insurance companies increased premiums by an average of $104.62 per person. During that same period, drug costs rose by $22.48.

Still, it's true that a majority of Americans with private health insurance are spending more for drugs—not only because they're taking more medication but also because their insurance is paying less. And it's no surprise that with rising pharmacy co-pays—the only healthcare costs that many of us actually see and feel—we tend to swallow the lie that increased healthcare costs are Big Pharma's fault.

Should we blame "Big Insurance"? Out-of-control out-of-pocket expenses cause many patients to stop using prescription drugs for controllable chronic conditions. The unfortunate result is that visits to the emergency room have jumped by 17 percent and hospital stays have risen 10 percent. And a new Integrated Benefits Institute study shows that when employers shift too much of their healthcare costs to employees, the companies lose more than they save, through absenteeism and lost productivity.

Should we blame our skewed priorities? American healthcare often works miracles when people become very ill, but it needs to do a better job with preventive care. Equally to blame is the fact that we spend a disproportionate amount of our healthcare budget on end-of-life care.

But rather than tangle up the already volatile healthcare debate in ethical arguments over whose life is worth more, it would be smarter to shift the focus to keeping people healthier longer. Earlier diagnosis and care are crucial to the future health of both Americans and American healthcare—and the pharmaceutical industry has a starring role here.

We cannot afford, in terms of dollars or lives, to continue the blame game. In order to deliver on the promise of affordable and quality healthcare for all citizens, all the players in the healthcare debate must work together. At the end of the day, we should unite against our common enemy—disease. The complicated and conflicting dynamics of politics, perspectives on healthcare economics, friction between payers, providers, manufacturers, and regulators, the battle for better patient education, and the need for a more forceful and factual debate over the value of innovation all require a more balanced and robust conversation.

Drug Development in the United States

Where do medicines come from? This may sound like a simple and straightforward question—and it is. New medicines come from the hard work of scientists who spend their professional lives looking for ways to improve the human condition. They work for the pharmaceutical industry, academia, and government, and their work largely takes place within the borders of the United States. Why? Because the United States rewards value and recognizes the importance of patents and intellectual property protection. As Abraham Lincoln said, the American system of patents adds "the fuel of interest to the fire of genius in the discovery and production of new and useful things." But there's danger lurking, and its name is politics. Today, many of the policy ideas being shared and shouted by politicians and pundits to "fix healthcare" would seriously undermine the basic principles of drug development and our nation's global leadership position. Attention must be paid.

Debunking the Myth of Me-Tooism

Originally published November 14, 2006

Almost exactly two years ago [9/30/04], one of the nation's largest pharmaceutical companies announced the voluntary worldwide withdrawal of Vioxx, the enormously popular arthritis drug.

Last month, two papers in the *Journal of the American Medical Association* examined the heart risks posed by Vioxx and concluded that the drug may have posed unique heart risks not shared by similar painkilling drugs, including Celebrex.

In other words—just as a Honda Accord isn't the same as a Toyota Camry— Celebrex isn't the same as Vioxx.

Common sense, right?

Not really.

Time and again, advocates of nationalized healthcare claim that there's no difference between so-called me-too drugs, and that they're simply a way for drug companies to turn a quick profit.

Leading this movement is Harvard Medical School professor Marcia Angell, who has achieved near-celebrity status by bashing drug companies. Author of the hugely successful book *The Truth About the Drug Companies: How They Deceive Us and What to Do About It*, Angell argues that in any given therapeutic class, only one or two drugs is needed.

"I know of no rationale," she wrote, "for, say, the seven brand-name ACE inhibitors that are sold to treat high blood pressure and heart failure."

And as the former editor in chief of the *New England Journal of Medicine*, Angell has done much to promulgate this notion. As she claimed in an interview with PBS, "the drug companies have found that the best way to make money at low cost is by turning out drugs that are imitations of other companies' blockbusters."

At that point, according to Angell, it's only a matter of promotion. "You have to have a huge marketing budget to convince the public that Nexium is better than Prilosec."

Accordingly, Angell's allies—including many in mainstream media—claim that in any given class of drugs, most medicines are interchangeable.

Just last year, for instance, in an editorial discussing drugs that treat schizophrenia, the *New York Times* claimed that "the nation is wasting billions of dollars on heavily marketed drugs that have never proved themselves in head-to-head competition against cheaper competitors."

That's why advocates of greater federal intervention in the drug market endorse restrictive government formularies, which limit a physician's ability to prescribe more than one drug in the same class.

When it comes to cholesterol-lowering pills, for example, this would mean that Lipitor, Crestor, Zocor, Pravachol, and Lescol are exactly the same—even though some pills are used to treat mild cholesterol elevations, others are used for severely elevated cholesterol, and still others are prescribed because they're not as likely to interact with certain medications.

The truth is that despite the assertions made by Angell and the *New York Times*, different drugs are indeed different, even if you describe them as me-too medications. Likewise, even though every brand of store-bought peanut butter contains peanuts, oil, sugar, and salt, no one would argue that Skippy and Jif taste exactly the same.

Time and again, different medicines have proved themselves where it counts—in the bodies and biochemistries of patients undergoing treatment. No two patients are alike, and the more options available, the more likely doctors are to find what works best for each patient.

If the Angells of the world had their way, there might not have been a safer alternative to Vioxx. People with arthritis would have had no options once the drug was pulled from the shelf.

Luckily for those with persistent joint pain, however, America's doctors, drug researchers, and patients are smarter than that. Just as there aren't any me-too patients, there aren't any me-too drugs. And if cars and peanut butter are going to be personalized, then medicine is too important not to be.

Four Bucks: Cheez Whiz and Imitation Prozac

Originally published November 22, 2006

What can you get for four bucks these days? Cheese in a spray can? Cinnamon dental floss? How about a month's supply of lifesaving prescription drugs?

Thanks to Wal-Mart, that's now possible.

Late last month, the company announced the expansion of its recently launched, $4 generic drug program to cover 27 states, or 2,507 pharmacies across the nation.

Contrary to popular wisdom, it would seem that markets are, in fact, capable of providing medicines at affordable prices.

Even in Vermont, where the boldest socialized healthcare dreams take wing, the *Rutland Herald* lauded Wal-Mart's initiative, editorializing that it "will make available drugs for all customers, including people without insurance," and that it "shows that even the pharmaceutical industry is susceptible to the market power of the major players."

So why is everyone surprised?

Americans demand affordable prescription drugs, and companies compete to supply them at a profit. The demand is so strong, in fact, that within four days of announcing its initial program in Florida, Wal-Mart had filled over 152,000 prescriptions for 314 different kinds of drugs.

Wal-Mart's customers, the company reports, are now enjoying extraordinary discounts on treatments for everything from Parkinson's and cancer to the common cold—on every medication from antibiotics to antipsychotics. Vitamins, too, so you don't have to be sick to benefit from it.

Just as predictably, Wal-Mart's competitors aren't allowing it to corner the market. Two days before Wal-Mart widened its $4 generics push, Medco Health Solutions, headquartered in Franklin Lakes, New Jersey, announced a new partnership with Nationwide, a Columbus, Ohio–based entity, to launch a competing service called Generics First.

For $10, Generics First provides a 90-day supply of any of 2,000 generic drugs to small businesses in 31 states and the District of Columbia. By nature, these businesses lack the numbers of employees necessary to pool risks and lower their insurance costs, so by making its purchasing power theirs, Medco has enabled them to buy medicines at dramatically discounted prices.

Target and Wegmans are also now on course to lower their prices.

Generic drugs, which anyone can produce once the patents and intellectual property rights held by a drug's creators expire, generally cost between a third and two-thirds less than brand-name products.

What makes generics so inexpensive is that companies compete to provide them ever more cheaply. And thanks to the absence of price regulation in America, they're allowed to.

Generics are cheaper in the United States than almost anywhere else in the developed world, including Canada. It doesn't make for exciting headlines, but many Canadians actually cross the border to buy our prescription drugs.

Wal-Mart estimates that under the $4 generics program, its customers will save $15.5 million a year on the antibiotic amoxicillin, and $23.2 million per year on fluoxetine, the generic form of the antidepressant Prozac. If that doesn't make you smile, what will?

Wal-Mart has also anticipated the need to coordinate customer payments with Medicare's Part D prescription drug benefit, so it has designated an employee in each of its pharmacies to ensure that customers receive the maximum benefit to which they are legally entitled.

The genius of competitive markets, for medicine and everything else, is that they allow individuals to decide what they need. And they operate on the assumption that one size doesn't fit all.

Everyone needs different kinds of medical coverage, different kinds of care, and different kinds of drugs. So different models spring up to satisfy the demand:

At Wal-Mart, $4 generics for individual customers.

At Generics First, $10 drugs for small businesses seeking to lower their medical costs.

Elsewhere? Who knows what's next.

In a country where people have enough disposable income to pay for a double soy peppermint Frappucino every day, we're right to expect a medicine for every malady. The market *can* satisfy the demand for affordable drugs.

Wal-Mart and its competitors are proving it.

Alzheimer's Setback Shows Difficulty of Drug Development

Originally published September 23, 2010

Pharmaceutical firm Eli Lilly recently announced that it would halt clinical tests for an experimental Alzheimer's treatment. The drug's failure was extremely disappointing, as it represented one of only five Alzheimer's drugs under development to have even reached late-stage clinical trials.

Lilly's announcement exemplifies the long odds that most drug researchers face in trying to identify cures for the world's most debilitating diseases.

What needs to be addressed are the twin issues of drug development and regulatory science. Both are lagging.

Too many drug trials—almost 50 percent—are failing in the Phase III trial stage, the latest stage, because they are mired in regulatory treacle. This is an unsustainable economic model from a research and development standpoint, and the impact of Alzheimer's on patients, their families, and American healthcare is devastating.

Better, more current and predictable scientific research and standards must be developed and devoted to streamlining the critical path. Investment in basic research is not enough. Specifically, new development tools, such as biomarkers, microarrays, and other diagnostic tools, are needed to improve the predictability of the drug-development cycle and to lower the cost of research by helping industry identify product failures earlier in the clinical trials process.

A quarter century ago, the success rate for a new drug used was about 14 percent. Today, a new medicinal compound entering Phase 1 testing—often after more than a decade of preclinical screening and evaluation—is estimated to have only an 8 percent chance of reaching the market. For very innovative and unproven technologies, the probability of a product's ability to make it to the market is even lower. We must work together to turn that around.

When Thomas Edison was asked why he was so successful, he responded, "Because I fail so much faster than everyone else." Consider the implications if the FDA could help companies fail faster. Using the lower end of the Tufts University estimate of the average pretax cost of new drug development ($802 million):

- A 10 percent improvement in predicting failure before clinical trials could save $100 million in development costs.

- Shifting 5 percent of clinical failures to Phase 1, the earliest stage, from Phase 3, the latest stage, reduces out-of-pocket costs for developers by $15–$20 million.

- Shifting of failures to Phase 1 from Phase 2, the middle stage, would reduce their out-of-pocket costs by $12–$21 million.

All of these dollars could then be reinvested in other innovative development programs for new lifesaving medicines.

For all that modern science has to offer, developing new treatments is still very much an art, in which hunches, intuition, and luck play a critical role. The odds are long. But for more medicine that is affordable and innovative, we need up-to-date regulations that complement the drug trial process in order to take these chances—which is precisely the mission of the FDA's moribund Reagan-Udall Foundation. The failure of Congress to free up the seed funding the Foundation has called for in the 2007 FDA Amendments Act must be corrected—if not in this Congress then in the next.

Senator Ted Kennedy said the Reagan-Udall Foundation "will make new research tools and techniques available to the entire research community, shortening the time it takes to develop new drugs and reducing costs for patients."

Let's hope he was right.

The High Cost of Slashing Patents

Originally published May 6, 2013

Medicine is getting cheaper. That may come as a surprise amid hand-wringing about the spiraling cost of healthcare, but two new studies, one from research company IMS Health and one from pharmacy benefit manager Express Scripts, show that the amount of money Americans spend on prescription drugs went down in 2012 for the first time in decades.

The reason for this welcome development is an influx of generic medications. Recently, the patents on a slew of blockbuster drugs—such as Lipitor, which fights cholesterol, and Plavix, which prevents blood clots—have expired, paving the way for less expensive versions. The research behind a new drug is protected for a fixed number of years, after which competing firms can begin manufacturing generic forms.

In 2012, 84 percent of all prescriptions were dispensed as generics, the highest rate in history. It's a boon for consumers.

These new studies also found the prices of specialty medicines are rising. These new drugs involve cutting-edge technologies and can, therefore, be expensive, priced as they are to help inventors recoup their investments. Fearful of what the newest medicines may cost patients and insurers, some politicians have proposed measures aimed at forcing these prices down.

We shouldn't fear the price tag of these new medicines. Expensive medicine may be a bitter pill, but these advanced therapies offer hope to millions of patients, keeping them healthier for longer. Lawmakers must continue to promote smart policies that encourage the research investments critical to invention.

We're living in a golden age of drug development. New treatments for everything from cancer to rare genetic diseases are entering the market all the time, many of which are cutting-edge biologic medicines derived from living cells.

Biologics offer amazing promise. Consider their potential impact on cancer. Conventional cancer treatments often generate significant collateral damage to the patient. In contrast, the biologic approach injects a genetically engineered protein designed to knock out a tumor's ability to produce new blood vessels, thereby cutting off its capacity to grow. No innocent tissue is harmed in the process. Such a biologic has already been approved for treating colorectal cancer.

Or consider a vaccine that, when injected directly into a tumor, would not only destroy the malignant cells but also stimulate the body's immune system to go after similar tumor cells. That therapy for treating melanoma is already in the development pipeline, along with 906 other biologics targeting over 100 diseases from autoimmune disorders to viruses. There are currently 176 biologics in development to treat infectious diseases alone.

But the most specialized and complex drugs can come at an astronomical price. According to an exclusive *Forbes* survey of the most expensive medications, four biopharmaceuticals approved in 2012 cost more than $200,000 per year, per patient.

That's because it costs on average, $1.2 billion to bring a new drug to market—from the time it is a twinkle in a scientist's eye, through a decade or more of lab research, to clinical trials, and finally FDA approval. To put this in perspective, the entire cost of one of the greatest physics experiments of all time, the search for the Higgs Boson—including building a super collider—would produce no more than 10 drugs.

The beauty of our system is that it encourages companies to make the massive investments of time and money required to bring a new drug to market. The biopharmaceutical industry's legacy of risk-taking research has led to a world in which eight of every ten medicines dispensed is generic.

And study after study proves these investments are paying off. Innovative therapies, though costly, are far more effective. By treating patients faster and keeping them well, advanced pharmaceuticals lower healthcare costs elsewhere.

For example, researchers are working to develop a medicine to delay the onset of Alzheimer's—the sixth leading cause of death in the United States. Such a breakthrough therapy could reduce the cost of care for Alzheimer's patients in 2050 by $447 billion.

Our children and grandchildren will grow up to marvel at the biologic revolution, just as an earlier generation marveled at the space race. But that can only happen if we accept the reality that innovation comes at a high price.

Federal Funds Should Go to Medicine-Makers

Originally published May 11, 2014

Senator Richard Durbin, D-Ill., this year proposed that the U.S. spend $150 billion more over 10 years on biomedical research at the National Institutes of Health (NIH) and other institutions.

Is more federal funding for the NIH the best bang for the buck when it comes to using precious tax dollars to advance public health? No.

The NIH budget is about $30 billion. But what does that buy? Where do discoveries that advance public health really come from? Some do come from NIH-funded research—but not nearly the majority. The engine of innovation is the biopharmaceutical industry, which spends in excess of $50 billion annually on research and development. It's not a competition; the NIH and industry complement each other's efforts. But context matters.

The NIH focuses on basic research, the study of fundamental aspects of phenomena without specific applications. The biopharmaceutical industry addresses most of its research and development toward clinical research, science focused on the actual development of new medicines. The NIH provides grants to academic institutions. Industry employs the scientists who do the work and, increasingly, funds academic research.

But there's a problem. While universities love NIH dollars, they are less enamored of industry resources. Why? One reason is that NIH funding counts toward achieving tenure, while similar dollars from biopharmaceutical companies do not. Durbin's legislation would disincentivize more industry-academic partnerships. More government spending is not always the mother of invention.

As the Prairie State's great Senator Everett Dirksen once (allegedly) quipped, "A billion here, a billion there, and pretty soon you're talking about real money." Some $10 billion annually could be allocated elsewhere to achieve broader access to newer, more targeted therapeutic medicines (and $5 billion could go toward the NIH's good work, hardly a paltry sum). The Food and Drug Administration should be No. 1 on the list to get more money.

The FDA regulates more than 25 percent of the U.S. economy, yet operates on an annual budget of $4.7 billion (about $2 billion generated by industry user fees). The budget's federal funding portion is about one-tenth the NIH's. Why hasn't Durbin proposed additional tax dollars for the FDA's programs on advancing regulatory science, expedited review pathways, or more-ready access to experimental medicines for desperately ill patients? The FDA doesn't even need $10 billion a year for 10 years to become our nation's leading force in healthcare innovation. Some $1 billion a year would do the job quite nicely. As to the remaining $9 billion, the line forms to the left.

Alas, headlines for hyped and misleading "NIH-funded cures" are far sexier than those for "more money for drug regulation." They may not be inversely important, but they are equally urgent in advancing 21st-century American healthcare.

Disarming the Patent Death Squad

Originally published June 11, 2015

When Ultratec, a manufacturer of closed-captioned phones for the deaf, realized that a rival had created a knockoff using its patented technology, the company filed a patent-infringement lawsuit. A Wisconsin federal jury ruled in Ultratec's favor in October 2014, ordering the rival, Sorenson Communications, to pay $44 million in damages.

But Ultratec may never receive a cent. In March 2015 a little-known but hugely powerful federal body called the Patent Trial and Appeal Board (PTAB) invalidated Ultratec's patents, on the grounds that the designs were too obvious to be patentable.

The PTAB, created by the 2011 America Invents Act, was intended to strengthen the patent system. Lawmakers hoped to avoid the need for patent lawsuits by giving patent holders and challengers an alternative to the courts, a quick and inexpensive way to resolve disputes.

But the board uses looser standards than a federal court to evaluate a patent's legitimacy. Courts assume that a patent is valid until a challenger provides "clear and convincing" evidence to the contrary. The PTAB requires only that challengers show that it's more likely than not (i.e., a "preponderance of the evidence") that a patent is too broad.

In recent months the board has overturned patents on a computer memory technology, a popular video game, and a system for monitoring car tires. The PTAB has invalidated at least one "claim"—or part—in almost 80 percent of the patents it has ruled on, according to a study in the *University of Chicago Law Review*. Some patent experts such as Randall Rader, former chief judge at the U.S. Court of Appeals for the Federal Circuit, have referred to the 300-odd administrative judges, attorneys, and legal aids on the board as "patent death squads."

Patent challengers have jumped at the chance to exploit the board's lax standards. Since it began to operate in September 2012, the PTAB has received over 2,600 patent challenge requests—three times more than it expected.

Many of these challenges—such as one against Combigan, an eye-drop medicine that prevents blindness in patients with glaucoma—seek to overturn patents that district courts have already upheld. In many other cases, the patents have also been challenged in federal courts—but courts have stayed the litigation until the PTAB has ruled. The patent may be invalidated without facing a court's stricter standard.

The PTAB could devastate innovation-intensive industries. Consider pharmaceutical developers, which spend about $51 billion a year researching new treatments. But less than 12 percent of drugs that reach clinical trials ever make it to market.

Patents give firms the financial incentive to fund challenging research and development projects. In the event that a huge upfront investment actually results in a popular new product, the developer can recoup its costs in sales.

The PTAB jeopardizes this process. Since an overturned patent means that rival companies could create knockoff products, firms will lose the confidence that they'll reap the rewards of innovation.

Some financiers have started using the PTAB to make a quick buck. Kyle Bass is a hedge-fund manager, not a pharmaceutical developer, but he recently challenged six drug patents. His strategy, which has been widely reported including in the *Wall Street Journal*, is to bet that the challenges would drive down the patent owners' stock prices.

The strategy is working. Early this year Mr. Bass challenged Acorda Therapeutics' patent on Ampyra, a medicine that uses a re-engineered bird poison to help multiple sclerosis patients walk. The claim: Medical experts would have been able to deduce the effectiveness and proper dosing of the re-engineered molecule. The challenge caused the company's stock price to drop almost 10 percent.

If hedge funds and copycats continue to take advantage of the PTAB's bias against patent holders, it will choke off funding for lifesaving medicines.

The PTAB will make it harder to create the products that improve lives and fuel the economy. To avoid this dangerous outcome, Congress has to reform the PTAB so that it operates under the same standards as a regular court.

360° Trade

Originally published June 13, 2016

When it comes to global trade, there is no value to living in the past. We no longer trade in barter or beads. The Trans-Pacific Partnership (TPP) acknowledges that we live in a global village—and innovation is an essential commodity. Why are some people finding this so surprising?

The evolution of the TPP has been long and arduous—but the basic premise has never changed. In order for global trade to flourish in an equitable manner, there have to be rules. Minus the rule of law, chaos ensues and we find ourselves in a survival of the fittest situation without fairness, predictability, or incentives for innovation. In a world without rules, ambiguity trumps investment and risk outweighs rewards.

One of the more important aspects of the TPP is intellectual property (IP) protection for pharmaceutical innovation. This is of particular importance to the United States, where the U.S. biopharmaceutical research sector leads the world in the development of new medicines with about 4,000 in development or FDA review in the U.S. and more than 7,000 in development worldwide. This sector generates high-quality jobs and powers economic output and exports for the U.S. economy, serving as the foundation upon which one of the U.S.'s most dynamic innovation and business ecosystems is built.

According to the Information Technology and Innovation Foundation (ITIF):

> America's biopharmaceutical companies are among its most innovative and commercially important. In 2014, the sector generated $97 billion in economic value-added, produced $54 billion in exports, and supported more than 3.4 million jobs. As measured by Battelle, the overall economic impact of the biopharmaceutical sector on the U.S. economy totals $789 billion on an annual basis when direct, indirect, and induced effects are considered. Moreover, the sector is extremely research-intensive, investing over 21 percent of its sales in research and development (R&D), while accounting for 23 percent of domestic R&D funded by U.S. businesses—more than any other sector. And measured by R&D expenditure per employee, the U.S. biopharmaceutical sector leads all other U.S. manufacturing sectors, investing more than 10 times the amount of R&D per employee than the average U.S. manufacturing sector. Strong private and public sector investment has made the

United States the world's largest global funder of biomedical R&D investment over the past two decades, a share that some analyses suggested reached as high as 70 to 80 percent.

Let's not forget the wise words of our sixteenth president, Abraham Lincoln, who commented that patents "add the fuel of interest to the passion of genius."

The TPP's life sciences IP provisions make progress in several important areas toward creating a robust regional innovation ecosystem. While some nations already had data protection obligations in place, the TPP commits countries to provide patent term adjustments for unreasonable curtailments of effective patent terms. It includes measures improving transparency in the listing and drug reimbursement programs run by national healthcare authorities. And it commits countries—such as Vietnam—that had previously lacked explicit data protection periods for the clinical trial data of biologic drugs to introduce them.

Data exclusivity protects the investment needed to develop the necessary clinical data, not the initial innovation. Patent protection rewards innovation and disclosure of that innovation to the public. Data exclusivity protects the investment necessary to generate the extensive clinical and other data that are necessary for, but that do not guarantee, FDA approval. It also encourages further research and development following initial product approval—research and development that has led to medical advances in treatments and patient care. Data exclusivity does not prevent competitors from entering the market, if those competitors generate their own safety and efficacy data.

It is essential to provide innovative manufacturers with a period of exclusive use of the data they generated, in order to help recoup the investment necessary to develop data, and to encourage future research and development.

If American innovation isn't protected, not only will we not, as a nation, benefit economically, but also the world will suffer the unfortunate unintended consequence of no innovation. And that is not acceptable.

The United States, with a largely market-based system, rewards the major risks that must be taken to bring new drugs to market. New drug development cannot occur unless innovators have the opportunity to be compensated for their financial risks. Put simply, the new medicines of today allow our industry to continue research into the cures of tomorrow. Our IP system, which covers patents and data protection, is among the strongest in the world, which is why the U.S. has the most medicines in development.

However, our system also encourages competition from generics and biosimilar manufacturers. This is of great benefit to U.S. patients: new medicines improve patient lives and can help reduce healthcare spending by mitigating the need for costly surgery or hospital visits, while allowing follow-on manufacturers to profit and compete once those IP protections expire. Generics and biosimilars would not exist without the IP of the innovators who developed them and assumed all the risks and costs associated with bringing a new drug to market.

In fact, a healthy innovative biopharmaceutical sector is a prerequisite for the generics and biosimilar industries to thrive. In short, for the generics and biosimilar industries to continue their growth, IP rights must continue to help spawn the medicines of tomorrow.

It is not by accident that pharmaceutical innovation is a key point in the debate over the TPP.

Per the ITIF report:

> Congressional Trade Promotion Authority directed the Obama administration's trade negotiators to seek IP protections similar to those enshrined in U.S. law. Thus—while certainly achieving progress with regard to nations that previously lacked biologics data protection altogether—it is disappointing that the TPP commits partners to provide at most eight years of data exclusivity protection.

But even if all TPP partners were to clearly enact eight years of data exclusivity protection for biologic drugs (and outside of Canada and Japan, which already provide eight years of regulatory data protection for biologics, this is far from a certainty), the TPP will still have fallen short of promoting globally a 12-year data exclusivity standard that has proven instrumental in contributing to world-leading levels of biomedical innovation being produced from the United States.

This represents a step back compared to the only other regional group of nations to have established a biologics data exclusivity standard—the European Union, with at least 10 years of data protection—thus setting a lower global standard for data exclusivity protections for biologics. This matters significantly, not only with regard to the countries currently participating in the TPP, but also to countries that may join the TPP in the future—such as China, Indonesia, or Korea. With as much as half of U.S. pharmaceutical companies' revenues now stemming from foreign sales, the TPP's eight-year data exclusivity standard will constrain some share of those revenues, relative to a 12-year standard.

Appropriately robust IP protection doesn't only benefit the United States but encourages faster entry of innovative medicines into overseas markets. In addition, growth in generics and maintaining the new drug pipeline are not mutually exclusive, as demonstrated in the United States where there is not only a strong IP system that supports innovation, but where generics penetration is at 88 percent. Further, there is no evidence that watering down IP protections has in any way helped tackle the real access challenges developing economies face; however, it is clear that a lack of commitment to protect IP in trade agreements will ultimately impair future R&D necessary to help patients, grow innovative ecosystems, and develop the next generation of therapies.

Failure to take full advantage of the opportunity provided by the TPP to establish a 12-year standard of regulatory data protection can only limit the promise and potential of biologics.

To borrow an overused adjective from the world of global climate change, we must protect "sustainable" innovation.

A Flawed Study Depicts Drug Companies as Profiteers

Originally published October 9, 2017

Are drug companies ripping off cancer patients? Of course they are, suggests a much-hyped study published last month in the journal *JAMA Internal Medicine*. Yet the truth behind the hype is more complicated.

Drug companies receive a staggering return on investment "not seen in other sectors of the economy," write Vinay Prasad of Oregon Health and Science University and Sham Mailankody of Memorial Sloan Kettering Cancer Center. They estimate that, on average, pharmaceutical firms spend $720 million to develop a single cancer drug, while the average cancer therapy generates sales of $6.7 billion.

The editors at JAMA are brilliant physicians, but they could use a refresher on the economics of drug development. Several methodological flaws in the study led the authors to underestimate drug-development costs.

Messrs. Prasad and Mailankody examined 10 publicly traded companies that secured their first-ever Food and Drug Administration approval between 2006 and 2015. They pulled data on companies' research spending and revenues from annual financial reports filed with the Securities and Exchange Commission. These selection criteria are a joke. By looking at 10 companies that only produced one cancer drug each, the authors screened out big multinational companies that had previously secured FDA approval for one or more drugs. Small biotech firms that hadn't secured FDA approval for any treatments were also excluded.

The authors admit that the selection criteria are a "critical limitation." No kidding: They only looked at 15 percent of all cancer drugs approved from 2006 to 2015, ignoring the other 85 percent of cancer therapies introduced that decade. This helped them "prove" their hypothesis.

The analysis also overlooks hundreds of millions of dollars of research spending at companies that never develop an FDA-approved medicine. Nine of every 10 publicly traded drug companies failed to break even in 2014, according to a 2016 International Trade Administration report. Most therapies don't make it out of the lab and into clinical trials. Of those that do, only 12 percent are brought to market.

Those that defy the odds and win FDA approval don't accurately represent the broader biopharmaceutical industry. Consider the success of these 10 drugs against those that are still going through clinical trials. Even if all of these companies' other experimental drugs in the development pipeline failed, the success of this study's 10 drugs would have resulted in an overall clinical approval success rate of 23 percent, twice the industry average.

Worse, five of the companies in question had purchased their drugs from smaller biotech firms. The authors didn't count any of the research and development expenditures of these "nurturer" firms, only spending by the acquiring firm.

The other five drugs were developed entirely in-house—and the authors lowballed cost estimates for developing these drugs. Messrs. Prasad and Mailankody counted only two years of development costs before the first mention of the drugs in the medical literature. They figured this would accurately reflect preclinical costs, such as lab tests.

Their assumption is wrong. In reality, the initial, preclinical research period often lasts four years or more. And for four of the ten drugs examined, companies started lab research at least seven years before the first mention of the drug in any published medical studies.

Drug development is much more expensive than the JAMA study suggests. More reliable is a November 2014 study from the Tufts Center for the Study of Drug Development. This more thorough examination estimates total research costs are about $2.6 billion for new cancer drugs. Politicians who advocate price controls undoubtedly will cite the JAMA study anyway. Never mind that government-imposed price caps would hamstring researchers and prevent the development of new treatments and cures.

In the past 17 years, biopharmaceutical companies have invented more than 550 FDA-approved medications. More than 800 experimental cancer drugs are currently under development at companies of all sizes, from tiny biotechs to giant multinationals. By misinforming readers, the JAMA study undermined the great work that drug companies are doing today.

Elizabeth Warren's Threat to Medical Progress

Originally published January 6, 2019

"Everyone is entitled to his own opinion, but not to his own facts." —Daniel Patrick Moynihan

Politicians on the hunt for bigger and bolder ways to lower drug prices are taking aim at a far more central part of pharma's monopoly power: the patents the industry holds on its drugs.

Patents save lives and enhance the value of medicines. As Abraham Lincoln said, "Patents add the fuel of interest to the passion of genius." Two potential presidential aspirants are leading the charge: Senators Elizabeth Warren and Bernie Sanders.

Senator Elizabeth Warren (D-MA) mistakenly believes that pharmaceutical innovation is primarily driven by the National Institutes of Health (NIH). She's calling for aggressive use of the Bayh-Dole Act to use "march in" control prices on government inventions and has drafted legislation—called the Medical Innovation Act—that would strap the private sector with a big new fee.

A study in Health Affairs by Bhaven N. Sampat and Frank R. Lichtenberg puts the issue in a data-driven perspective that gives the NIH its due but in the proper frame of reference. Per Sampat and Lichtenberg, less than 10 percent of drugs had a public-sector patent. Drugs with public-sector patents accounted for 2.5 percent of sales, but the indirect impact was higher for drugs granted priority review by the FDA. Priority review is "given to drugs that offer major advances in treatment or provide a treatment where no adequate therapy exists."

"[Four hundred seventy-eight] drugs in our sample were associated with $132.7 billion in prescription drug sales in 2006. Drugs with public-sector patents accounted for 2.5 percent of these sales, while drugs whose applications cited federally funded research and development or government publications accounted for 27 percent," they wrote.

The NIH plays a vital role in basic research and early discovery, but is robbing Productive Peter to pay Government Paul the best bang for the buck when it comes to advancing public health?

The answer is a clear no. The primary engine of drug innovation is private industry. The members of the Pharmaceutical Research and Manufacturers

Association (PhRMA) spend in excess of $70 billion annually on research and development—and these are only some of the larger investors.

The NIH focuses on basic research—that is, the study of fundamental aspects of organic phenomena without regard to specific medical applications. The biopharmaceutical industry, on the other hand, directs most of its R&D toward clinical research. Private science is centered on the actual development of new medicines. If the government wants to get paid for the success of every molecule that comes from NIH basic science, should they be on the hook for every failure generated due to that science?

Both the NIH and private firms provide research financing to academic institutions. But it is industry that employs most of the scientists who conduct the hands-on development work. Unfortunately, some lawmakers have bought the myth that the NIH is primarily responsible for new medicines.

Pursuing misguided policies that siphon funding from groundbreaking medical research will have devastating consequences for patients and society. The proposed legislation would result in fewer medicines for patients and lost jobs at a time when our economy can least afford it. Senator Warren and others should pay heed to the facts and avoid the fiction. They are inversely important to advancing 21st-century healthcare.

Also on the hunt is Senator Bernie Sanders (I-VT), and it's not his first joust with pharmaceutical patents. In the past, he's introduced a bill that would replace our current patent system for pharmaceuticals with a "Medical Innovation Prize Fund."

It's not a new idea. The prize model has been used in the past by the old Soviet Union—and it didn't work. The Soviet experience was characterized by low levels of monetary compensation and poor innovative performance. The U.S. experience isn't much better. The federal government paid Robert Goddard (the father of American rocketry) $1 million as compensation for his basic liquid rocket patents. A fair price? Not when you consider that during the remaining life of those patents, U.S. expenditures on liquid-propelled rockets amounted to around $10 billion.

It's certainly not what the economist Joseph Schumpeter had in mind when he wrote about a "spectacular prize thrown to a small minority of winners." There's a difference between "creative destruction" and destroying medical innovation.

Senator Sanders's legislation would replace a patent system that has allowed the average American lifespan to increase by almost a full decade over the last 50 years with a prize program that has a solid record of complete failure.

As Joe DiMasi (Tufts University) and Henry Grabowski (Duke University) have argued, under a prize program, pharmaceutical innovators would lack the incentive to innovate. To quote DiMasi and Grabowski, "The dynamic benefits created by patents on pharmaceuticals can, and almost surely do, swamp in significance their short-run inefficiencies."

In other words (and to paraphrase Winston Churchill), our pharmaceutical patent system is the worst way to stimulate and support healthcare innovation except for every other system. A prize program shouldn't even appear on a list of 100 ideas for ways to improve innovation and access.

Who could support such a crackpot idea? Nobody? Wrong! As DiMasi and Grabowski presciently observed in 2004, "The main beneficiaries in the short-term would be private insurers and public sector purchaser of pharmaceuticals. Governments and insurers are focused myopically on managing health care costs. They are not likely to be strong advocates for funding new drug development that can increase individual quality of life and productivity."

At a time when we are finally focusing on the role of the middleman (payers, PBMs, etc.) now is precisely the time to focus on the *cui bono* of the healthcare ecosystem.

A prize in every box does not a crackerjack idea make.

As we move forward into a new year, a new Congress, and a presidential election cycle, there will be a lot of healthcare hyperbole. We need to be wary of populist ideas that are unworkable, ill-considered, and precarious—but that are rich in sexy soundbites.

Nulla mensa sine impensa: there is no free lunch.

Greedy Tort Bar Tarts Up the CREATES Act

Originally published February 12, 2019

When members of the tort bar start to salivate over a piece of legislation, it's worthwhile to find out where the red meat resides.

In a rush to pass legislation to "lower drug prices," lawmakers are pushing forward H.R. 965 (Cicilline and Sensenbrenner) and S.340 (Leahy). The worthwhile goal of this new version of the CREATES Act is to prohibit pharmaceutical and biologic companies from engaging in anticompetitive conduct that blocks lower-cost generic drugs from entering the market.

Both bills address an important problem—but create an even bigger one.

This proposed legislation establishes a private right of action for "eligible product developers" to sue "license holders" for failure to comply with the process set forth in CREATES for providing access to a covered product. That's good!

But, as written, the CREATES Act is ripe for abuse by entities that have no intent to actually develop a generic or biosimilar version of the covered product. This potential for abuse is exacerbated by the significant monetary damages available under CREATES—up to the amount of revenue generated on the covered product during the period of violation. Indeed, in certain instances, it may be more profitable to litigate and obtain damages under CREATES than it would be to actually market a generic or biosimilar product.

Can you hear the tort bar drooling? Can you hear the tort bar … drafting?

As currently drafted, the CREATES Act could have significant unintended consequences:

The damage provisions of the current draft create the potential for windfall damages, which will distort incentives. Specifically, the bill allows for damages up to the entire gross profit of the brand medicine during the period of negotiations. This creates a powerful incentive for generic companies to prolong negotiations (increasing their damage award), which will actually delay generic entry and competition in the marketplace.

Indeed, generic developers would be able to earn more from a lawsuit than from actually selling the proposed generic drug. An "opportunistic" company (can you say "Shkreli"?) could develop a business model of

34

demanding samples and engaging in litigation for damages without ever submitting an abbreviated application to FDA—undermining the bill's stated goal to speed availability of lower-cost drugs for patients.

Smart changes will fix these problems, save the government money, and achieve the desired policy goals of facilitating generic access to samples and increasing competition; maintaining safeguards for products subject to a REMS; and ensuring generic developers actually develop generic products.

Here's a savvy path forward:

Establish an affirmative defense for license holders where the license holder has made a timely offer to provide sufficient quantities of samples at commercially reasonable, market-based terms.

Such an affirmative defense is intended to prevent frivolous litigation—where samples are offered on commercially reasonable terms, the eligible product developer should not be able to decline the offer and continue litigation. The term "commercially reasonable, market-based terms" is defined to provide further clarity to all parties and avoid unnecessary litigation. Remedies would not be available if the license holder has established an affirmative defense by a "preponderance of the evidence." This affirmative defense also protects good acting companies from protracted litigation and stops generic sponsors from unnecessarily prolonging negotiations to increase damages.

Revise the definition of "eligible product developer" to clarify that the eligible product developer must be a person who seeks to develop "and submit" an application for a generic or biosimilar product.

These change will help ensure that only legitimate manufacturers are considered eligible product developers for purposes of CREATES; entities that seek only to engage in frivolous litigation and do not seek to submit a generic/biosimilar product application would not be eligible for the remedies under CREATES. The tort bar members won't like it, but this important legislation must be about lowering drug prices—not raising their income.

Cornyn-Blumenthal Drug Bill Carries Serious Side Effects, No Benefits

Originally Published May 24, 2019

Senators John Cornyn (R-TX) and Richard Blumenthal (D-CT) claim that their new bill—the Affordable Prescriptions for Patients Act—will expand access to needed medicines for American patients. But in reality, the bill is poised to do the exact opposite.

Senators Cornyn and Blumenthal want to target the drug industry for "anti-competitive abuses of the patent system." The chief consequence of their reform, however, will be to deter pharmaceutical firms from investing in new drugs, or even improving on existing medicines.

American patients will suffer in the process.

The bill is premised on the idea that drug companies are gaming the intellectual property (IP) system by making incremental changes to drugs they have already patented. As Senators Cornyn and Blumenthal see things, such behavior is merely a way of driving off competition from less expensive generics, thereby inflating drug costs for patients.

The Affordable Prescriptions for Patients Act addresses this perceived abuse by empowering the Federal Trade Commission (FTC) to bring antitrust suits against pharmaceutical companies that engage in this practice.

But, plainly speaking, nearly every assumption behind the bill is fundamentally flawed. For starters, the reform takes for granted that any alterations or refinements to an existing medicine automatically amount to an anti-competitive ploy, when in fact there are countless reasons to engage in such work—starting with better treatments for patients.

For instance, a drug firm might develop a time-release version of a medicine to make it easier and safer for patients to take. Or it might create new dosage forms entirely, allowing patients to take a pill or wear a patch instead of having the medicine administered intravenously. Similarly, researchers could find a new combination therapy that involves taking multiple already-available drugs in a specific sequence.

Advances like these can drastically benefit patients, whether by improving outcomes, reducing side effects, or simply making a drug regimen more convenient.

Consider the difference between traveling to a doctor's office to receive IV chemotherapy and taking a chemotherapy pill at home. For a patient who lives far away from a medical facility, or has trouble moving around, this new dosage form can be life-changing. And it's only reasonable that drug companies that invest in such improvement are able to patent their discoveries.

A bill that penalizes such work will ensure that legitimate enhancements of existing products aren't pursued—and patients will be worse off as a result.

It's also not clear how developing a new version of an old drug impedes access to generics. After all, these improved medicines don't extend the patent on the original drug. Any firm looking to release a lower-cost generic of a drug's earlier version is free to do so, as many in fact do.

Indeed, though Senators Cornyn and Blumenthal paint brand-name drug companies as monopolists, the fact is that roughly 80 percent of prescriptions filled in the United States are generics. So the notion that generic drugmakers are being unfairly kept out of the market simply doesn't comport with the facts.

What's most troubling about the Cornyn-Blumenthal reform, however, is the threat it poses to America's patent system, an institution that's essential to drug discovery.

It takes a decade or more to bring a single new drug to market, and costs, on average, $2.6 billion. That process almost always involves numerous false starts and dead ends—only one in ten drugs that reach clinical trials are ultimately approved.

Without the ability to protect their inventions through reliable IP laws, drug companies would have far less reason to expend the resources necessary to make genuine breakthroughs.

That would be disastrous for patients, especially those suffering from conditions for which no suitable treatment yet exists. Just think of those suffering from Alzheimer's, a disease which afflicts 10 percent of Americans over 65, and for which a new drug hasn't been approved in more than 15 years.

The Affordable Prescriptions for Patients Act would do little to expand access to generic drugs. But it would undermine an IP system that makes medical progress possible in the first place, leaving millions much worse off as a result. What Senators Cornyn and Blumenthal need is an innovative dose of reality.

CHAPTER TWO
Drug Importation

H.L. Mencken reminds us that, "For every complex problem there is an answer that is clear, simple, and wrong." Drug pricing is one such complex problem. And drug importation is a simple solution that's as dangerous as it is wrong. But that hasn't stopped politicians of all parties from touting it as a quick path to lower costs. "Drugs from Canada" sounds like such a simple and straightforward solution. The facts speak differently and forcefully if only we allow ourselves to tune out the rhetoric and focus on reality. Importation schemes at the local and state levels have all been dismal failures. Costs do not come down and patient safety is jeopardized. As John Adams said, "Facts are pesky things."

Is Hezbollah Filling Your Prescription?

Originally published October 13, 2006

With Mexico less than 20 miles south of downtown San Diego, you'd probably have a pretty easy time getting your hands on cheap foreign medicine. But while people in the market for "lifestyle" drugs such as Viagra may indeed be cruising to Tijuana, those who need medicines for chronic conditions such as high cholesterol and diabetes probably wouldn't risk purchasing a knockoff that might be a sugar pill—or worse.

If they're looking for cheap foreign alternatives, they're far more likely to purchase their prescriptions from a "Canadian" internet site. Most people think that buying foreign drugs from Canada is completely safe. Even Governor Arnold Schwarzenegger has been petitioning Congress "to allow Americans to import safe prescription drugs." And even though they're misinformed about the safety issue, the practice is now easier than ever before.

This week [Oct. 9], U.S. customs agents stopped seizing prescription drugs imported through the mail from Canada. Even though the drugs are illegal, customs agents—under pressure from Congress—have decided to cease enforcing the law.

This latest development comes only days after Congress approved an amendment to the Homeland Security appropriations bill that would also prohibit U.S. Customs and Border Protection from seizing prescription drugs purchased at brick-and-mortar pharmacies in Canada.

Senator David Vitter (R-La.), who sponsored the amendment, has touted the measure as "a major victory in [the] fight to bring lower-cost prescription drugs to all Americans." Vitter and his allies believe that their measure will ensure access to safe and effective pharmaceuticals.

Unfortunately, Canadian drugs are hardly safer than the ones coming in from Mexico.

Supposedly, a Canadian pharmacy cannot fill an American prescription unless it's co-signed by a Canadian doctor. And most Canadian provinces have a provision requiring doctors to actually examine the patient before co-signing a prescription.

But those laws are rarely followed.

That's why, despite conventional wisdom, walking into a pharmacy in Windsor,

Ontario, and having your prescription filled by pharmacist is no guarantee that your drugs will be genuine.

Why? Because most Canadian pharmacies situated near the American border employ physicians who do nothing more than blithely sign their names without even bothering to see what—or to whom—they're prescribing.

Consider this: Last year, a doctor in Toronto was prosecuted for co-signing 24,212 prescriptions over six and a half months so that U.S. citizens could have their prescriptions filled by an Ontario pharmacy. In addition to never having seen the patients, he charged $10 per signature.

Nice work if you can get it.

Considering that, on average, this one Toronto-based physician co-signed more than 124 prescriptions each day, it's obvious that many Americans are already purchasing their pharmaceuticals in Canada. But that's no reason to exacerbate the problem. By blocking Customs and Border Patrol agents from doing their jobs, Vitter's amendment would open the floodgates to dangerous drugs.

"This [bill] really breaks the dam, and it shows that it's only a matter of time before we pass a full-blown reimportation bill," Vitter said.

According to a recent report issued under the signature of the U.S. Surgeon General, opening up our borders to drugs from Canada would result in an uncontrollable influx of untested, impure, expired, and counterfeit drugs from around the world.

Further, terrorists have already proven that they are adept at exploiting weakened chains of pharmaceutical custody.

In fact, according to the federal Joint Terrorism Task Force, a global terrorist ring with ties to Hezbollah is importing counterfeit drugs into the United States by way of Canada. It's a great money-maker for rogue regimes, too. The Congressional Research Service says that North Korea is now producing counterfeits to fund its military development.

The last thing we need is to encourage the illegal drug trade. Without legalized prescription drug importation, the number of counterfeit drug investigations has increased four-fold since the late 1990s. And now that Customs agents are no longer allowed to do their jobs, it's even easier for terrorists and counterfeiters to abuse the porous nature of America's borders.

It's hard to see how any of this will improve our "homeland security"—or the health of Californians.

A Poison Pill, Indeed

Originally published May 24, 2007

Through a last-minute legislative maneuver, the U.S. Senate recently blocked a measure that would have allowed U.S. consumers to purchase prescription drugs from abroad.

Afterward, Senator Bernie Sanders (I-VT) described the maneuver as a "poison pill" for drug importation.

Oddly enough, Senator Sanders's flip comment hits at the very heart of this issue. Despite the fact that many lawmakers have been advocating drug importation for years, it is precisely the threat of poisonous, counterfeit, and unregulated drugs flooding the U.S. that has kept such legislation from passing.

Consider, for example, the widespread notion that drug importation would simply open our medicine cabinets to shipments from countries such as Canada, Great Britain, and France.

Such an impression couldn't be further from the truth.

Because of open trade between the European Union's 27 member states, drugs purchased from Britain or France could have originated in Latvia, Malta, Cyprus, or elsewhere in the EU

Further, most large internet pharmacies in Canada admit to getting their drugs from Europe. So drugs purchased from our northern neighbor could have easily started off in those countries, as well.

This raises serious safety concerns.

The EU doesn't require drug importers to record the batch numbers of parallel importations, so if a particular batch of medicines originally intended for sale in Greece is recalled, tracing where the entire batch has gone is impossible. Elsewhere in the EU, it's not uncommon to find misstated dosage strength, tablets errantly replaced with capsules, and mislabeled expiration dates.

Also, the world's drug market is awash in counterfeits.

According to the World Health Organization, as much as 10 percent of today's global medicine supply chain is counterfeit. And the U.S. Food and Drug Administration has reported that in some countries, that figure jumps to 50 percent or higher.

Moreover, terrorists have already proven that they are adept at exploiting weakened chains of pharmaceutical custody. According to a recent report from the federal Joint Terrorism Task Force, a global terrorist ring with ties to Hezbollah is importing counterfeit drugs into the United States by way of Canada.

But not to worry, say supporters of drug importation. Any legislation that passes, they promise, will include a clause demanding the FDA ensure that imported drugs are safe, legal, and properly labeled.

Simply demanding such a guarantee, however, hardly ensures that the FDA actually has the ability to follow through.

Finally, drug importation leaves our nation open to the serious health risks posed by an unregulated drug marketplace.

Because of Canada's Drug and Pharmacies Regulation Act, it is illegal for a Canadian pharmacy to dispense a prescription drug without the written authorization of a Canadian doctor. To skirt the law, internet pharmacies pay Canadian doctors to sign prescriptions for American patients.

This is a recipe for disaster.

A few years ago, for instance, a Toronto-based physician was found guilty of co-signing 24,212 prescriptions over six and a half months so that U.S. citizens could have their prescriptions filled by an Ontario pharmacy.

Considering that, on average, this one physician signed more than 124 prescriptions each day, it's obvious that he didn't seriously evaluate the merits of each script he signed.

And whether it's a teenager who hopes to mix Viagra, Ritalin, and vodka for the party of his life or a senior citizen who mistakenly orders the wrong dosage of Lipitor, the consequences of such renegade action can be deadly.

Despite these facts, drug importation isn't dead. The AARP and its congressional allies continue to claim that drug importation is a surefire way to give seniors and other consumers immediate access to cheaper drugs, and in the coming weeks, the House of Representatives is expected to consider the issue.

Policymakers must confront the very serious business of ensuring that drugs entering our markets are legitimate and safe. Such behavior might not score political points, but it will certainly make certain that we don't open our borders to an uncontrollable influx of untested, impure, expired, and counterfeit drugs from around the world.

Paying the Ultimate Price for Cheap Pills

Originally published October 25, 2007

When was the last time you picked up a bottle of Bayer Aspirin and wondered whether it was authentic?

Never? Well, Bayer is a well-known brand. You were probably in a chain store, or at least on a popular website such as DrugStore.com. Plus, you don't need a prescription for aspirin.

But what if you wanted a drug that might be unavailable or cost too much in your area? What would you do if, alternatively, the government, your doctor, or your insurance company denied your request for something you thought could help you?

Like many others, Marcia Bergeron of Quadra Island in Canada was in just such a situation when she turned to the internet.

While she likely thought she was purchasing inexpensive medicine from a safe online supplier in Canada, authorities found that she had actually ordered pills produced in Southeast Asia from a website based in the Czech Republic. It was a costly mistake that killed her.

Toxicology tests found three well-known drugs in her system: alprazolam—more commonly recognized as the antianxiety drug Xanax; zolpidem—which most are acquainted with as the brand-name sleeping pill Ambien; and acetaminophen. Zolpidem is not available in Canada, so it's understandable why Bergeron turned to the internet to get the drug.

The number of additional substances found in toxicology tests, however, was astounding. Bergeron had high levels of "filler" materials such as aluminum, tin, and even arsenic in her system. Such materials are the hallmark of dangerous counterfeit drugs.

As more consumers head online for cheap drugs, it's all too likely that Bergeron's story could become common—especially because the world's drug market is already awash in counterfeits.

The World Health Organization estimates that up to 10 percent of drugs sold around the world are fake. In fact, a 2005 FDA operation at the New York, Miami, and Los Angeles airports found that 85 percent of drugs from so-called Canadian pharmacies didn't actually come from Canada.

Supporters of drug importation hope to ease concerns about counterfeits by mandating the thorough inspection of imported drugs. Unfortunately, the FDA will have a hard time making sure everything is safe, legal, and properly labeled because counterfeiting is so decentralized.

In the European Union, for example, regulations prohibit manufacturers from managing their supply chains. This results in a lucrative market for wholesalers, who can open a package of pills only to repackage and relabel it. Because the manufacturer has no say in the process, dosages and expiration dates are frequently incorrect, and the drugs themselves are also subject to tampering.

So a box of meds that looks like it came from Britain could easily have originated anywhere in the EU, including places such as Latvia, Malta, or the Czech Republic.

Fortunately, elderly Americans don't need to scour the internet for cheap drugs. Anyone who's 65 or older qualifies for the Medicare drug benefit. Although this program is only two years old, it is already providing subsidized coverage to around 24 million seniors. Its remarkable success is due to the fact that the drug benefit, unlike most government programs, is administered by private companies.

Yet many lawmakers continue to claim that drug importation is necessary to give seniors and other consumers access to cheaper drugs. If the tragic story of Marcia Bergeron has proven anything, it's that turning to a pharmacy supposedly based in Canada to fill one's medical needs can have disastrous consequences.

Saving money should take coupons, not lives. If only policymakers kept up on the news.

The GAO's Alarming Message

Originally published February 25, 2008

Last week, the Government Accountability Office (GAO) delivered an alarming message to a Congressional panel—that the Food and Drug Administration lacks the resources to ensure the safety of the nation's food, drugs, and medical devices.

As a result, Americans aren't being adequately shielded from serious health risks.

Congress has vowed to fix the FDA. It plans to boost its funding, which is a good start. But those efforts will be more than undermined if Congress legalizes drug importation, which it may try to do this month. Attempting to screen imported prescription drugs would stretch the FDA's abilities to their breaking point.

Just as the agency's workload has surged, its budget has shrunk. The FDA's congressional appropriations have declined 400 million in inflation-adjusted dollars in the last 14 years. But its responsibilities have increased substantially, thanks to over 100 congressional measures expanding its regulatory scope over the last two decades.

Those budget cuts translate to 1,311 fewer workers today overseeing an ever-expanding territory. Eighty percent of all drugs sold in the U.S. are currently manufactured abroad, but there are just 380 inspectors in the FDA's ranks—a decline of 28 percent since 2003.

Consequently, the FDA cannot inspect the factories of foreign drugmakers as frequently as it should. According to the GAO, only 7 percent of the over 3,000 foreign drug factories registered with the FDA are inspected in any given year—which means that more than 13 years could pass between inspections.

By contrast, domestic drugmakers receive at least one FDA inspection every two years.

The FDA's inability to monitor the 714 Chinese drugmakers subject to agency rules is especially worrisome. Just 13 of them—1.8 percent—were inspected in 2007, according to the GAO.

That's not comforting news for American patients. The World Health Organization estimates that 10 percent of the world's drug supply is counterfeit, and many, if not most, of these fakes are made in China.

For example, it was recently reported that a Chinese drug company manufactured an antileukemia drug that left nearly two hundred Chinese paralyzed or otherwise harmed. Ordinarily, this would not have raised headlines in America. But the Chinese firm in question was also the sole supplier of the abortion pill R.U.-486 to the U.S.

Given how few foreign inspections the FDA conducts—especially in China, where the risks are greatest—the likelihood of a medical calamity befalling American patients is far too high.

With these dangers in mind, how can Congress possibly believe drug importation would be safe? Foreign-made pills would flood America's borders.

That means even more foreign packages potentially containing fake or harmful drugs will enter the country without being inspected. The John F. Kennedy Airport Mail Facility receives approximately 40,000 foreign shipments thought to contain drugs each day. Only 500–700 of them are inspected. It makes no sense to expand inspection responsibilities when the FDA is unable to cover its current ones.

Approving the drug-importation bill might push the agency over the edge. In the words of its former chief counsel, the FDA is "barely hanging on by its fingertips."

Now is not the time to saddle it even further with the impossible task of adequately screening drug imports. The American people can't afford it.

We're Taking Your Medicine, Literally

Originally published March 11, 2008

Imagine that you are an inventor and the government steals your highly lucrative idea, without any warning. The next day, you are informed that the government plans to mass-produce your invention and give it away for free.

This is what happens, with increasing regularity, to the manufacturers of lifesaving medicines. And self-appointed public health activists the world over are urging other governments to follow suit.

The most recent example occurred last year in Brazil, when the government confiscated the patent of an antiretroviral treatment for HIV produced by Merck, a U.S. pharmaceutical firm.

Brazil's decision came shortly after a similar move in Thailand, where the military-appointed government issued these "compulsory licenses" to obtain two drugs.

The first, the HIV/AIDS drug Kaletra, is produced by the U.S.-based Abbott Laboratories. The second, the heart-disease drug Plavix, is manufactured by Sanofi Aventis of France and Bristol Myers Squibb of the United States. The Thai government had already issued a compulsory license for Merck's antiretroviral.

Thailand's behavior is hardly unique. Across the world, it has been going on for years.

Last year in India, the government passed an amendment denying patents to pharmaceuticals derived from "previous knowledge," a purposefully arbitrary phrase. Also last year, India's Ranbaxy Laboratories began offering a generic form of Pfizer's cholesterol-lowering Lipitor in Denmark, despite the fact that Pfizer still holds the patent. Although Ranbaxy's version was already available in India and some other emerging markets, Denmark is the first Western country to sell a copycat version of the drug.

Meanwhile, many Latin American countries have repeatedly threatened the use of patent theft to strong-arm pharmaceutical companies.

Even worse, U.S. lawmakers are piling on. Last year, 22 members of Congress signed a letter to the U.S. Trade Representative expressing their support for the Thai government's decision.

Such action is a slap in the face to the companies whose expensive investments in drug research and technology ensure that these lifesaving medicines exist in the first place.

Thankfully, however, it finally appears that those responsible for ensuring global health are taking notice of the detrimental effects such sweeping policies have on the world's poor.

Just three days after the Thai government's decision, World Health Organization (WHO) Director-General Margaret Chan laid out the key reason why communicable diseases remain such a large problem in poor countries. As she explained, "[the pharmaceutical industry] has little incentive to develop drugs and vaccines for markets that cannot pay."

In other words, such theft discourages innovation. Drug development is an enormously expensive, time-consuming venture requiring years of effort by teams of highly trained researchers.

In 2004 alone, according to the Government Accountability Office, the pharmaceutical industry spent $60 billion on research and development. Indeed, the average drug costs nearly $1 billion to develop.

If a company stands no chance of recouping even a portion of that investment, where is its incentive to tackle the many diseases that ravage the third world?

Western drug companies may make good scapegoats, but in Thailand alone, they have contributed tens of millions of dollars to family planning programs and training programs for nurses and doctors treating HIV/AIDS patients. And it's doubtful that the Thai junta has the resources or the know-how to create such lifesaving drugs on their own.

Furthermore, there is no guarantee that generic drugs produced overseas will even work. Quite often, copycat versions of patented drugs are manufactured in factories that do not meet WHO standards.

No one can question that activists and governments alike wish to combat these diseases in the most efficient way possible. But the greatest challenge in the world's poorest nations is healthcare infrastructure—not pharmaceutical patents.

In the face of these larger structural challenges, patent theft is simply a cop-out—and a deadly one at that.

Three Ways the Nation's Top Drug Official Can Keep Americans Safe

Originally published June 10, 2009

Congress just confirmed Dr. Margaret Hamburg as the commissioner of the Food and Drug Administration (FDA). She's got her work cut out for her. The FDA is responsible for policing a wide array of consumer goods, from foods and pharmaceuticals to cosmetics and medical devices.

Dr. Hamburg has vowed to enhance the FDA's focus toward improving public health. That's the right move. But the agency has limited resources, so she'll have to identify policies that can do the most for the greatest number of Americans. Here are three that should top her list.

First: honestly addressing foreign drug importation. Some of our elected representatives, including the president, have signaled that they'd like to allow importation as part of an effort to lower prescription drug prices. Bad idea.

If Dr. Hamburg is committed to protecting public health, she'll resist these efforts. Most foreign countries don't adhere to the same rigorous drug safety standards as the United States. So there's a danger that imported drug shipments will contain medicines that are substandard and even counterfeit.

In 2003, the FDA found that 16 percent of drugs shipped into America from Canada violated federal safety statutes. A year later, FDA investigators bought three popular drugs from a variety of online pharmacies claiming to be in Canada. Every one of the drugs failed tests for potency and purity.

The situation across the Atlantic is no better. Recently, the Irish drug enforcement agency announced that it had confiscated more than 1.7 million pounds of counterfeit and illegal drug packages.

And remember, there's no such thing as drug "reimportation." That's putting lipstick on a pig. This isn't about making available to Americans the "same" drugs being sold to our neighbors to the north. This is about drugs from a multitude of nations (most notably those of the European Union). That means a drug that is said to come "from Great Britain" could just as easily originate from places such as Greece, Portugal, Latvia, or Malta. And basic economics dictates that goods from lower-cost markets flow into higher-cost ones. (In fact, more than 20 percent of all prescription medicines sold in the UK are "parallel imported" from lower-cost nations within the EU.) These drugs, arbitraged through Canadian internet pharmacies and then resold to Americans, may be dubbed "reimported" but that's just plain misleading—because these off-the-back-of-the-truck medicines are not even legal for sale in Canada. So much for "the same drugs" as you can buy at a pharmacy in Toronto.

Dr. Hamburg should recognize that the FDA can't conceivably guarantee the safety of imported drugs and encourage Congress not to open the American market to dangerous counterfeit drugs.

The FDA should also encourage the development of 21st-century biologics, the cutting-edge pharmaceuticals derived from living proteins. Biologics have proven effective at treating some of the deadliest diseases, including multiple sclerosis, cancer, and AIDS. Future biologic innovation could lead to therapies—or even cures—for diseases now considered untreatable.

Congress is considering several bills that would allow for the approval of copies of biologics, or biosimilars. Like generic versions of traditional chemical drugs, biosimilars would be cheaper than their brand-name counterparts. But unlike conventional generics, biosimilars aren't exact copies of the drugs they mimic.

It's impossible to replicate a biologic because of the complexity of manufacturing something using living cells. So biosimilars often contain small—but important—chemical differences from the drugs they're based on.

Therefore, biosimilars should be required to undergo independent safety trials before hitting the market. One of the bills being considered by Congress, the "Pathway for Biosimilars Act," stipulates just that. In the interest of patient safety, Dr. Hamburg should support this measure.

Finally, the new FDA chief should continue her predecessor's commitment to the Critical Path Initiative. This five-year-old program aims to expedite the drug-development process. Too often, steps in the approval chain are so costly or confusing that it takes longer than necessary to turn scientific discoveries into marketable medicines.

The Critical Path Initiative employs technologies such as genomics to improve the tests used to predict the safety and efficacy of medical products.

The new FDA commissioner faces unprecedented challenges. But by focusing on these three issues, she can ensure that American citizens remain safe—and that they continue to benefit from modern medical research.

The Myth of Drug "Reimportation"

Originally published July 28, 2009

On Thursday, as part of the Department of Homeland Security funding bill, the Senate voted to make us less secure by allowing Americans to purchase prescription drugs from Canada over the internet. The measure is now headed to conference, where House and Senate lawmakers will hammer out a final piece of legislation.

When he introduced the measure to his fellow senators, Louisiana Republican David Vitter described it as a "reimportation amendment." And over the next few weeks, as lawmakers deliberate on this, you're likely to hear that phrase quite a bit. Supporters of foreign drug importation believe that such wording makes this policy more palatable to the American public.

After all, the implication of the term "reimportation" is that once the ban is lifted, U.S. manufacturers will export their drugs to foreign distributors, which will, in turn, sell back those exact same drugs to us.

Brand-name pharmaceuticals found abroad tend to be significantly cheaper than they are in the States, largely because foreign governments impose stringent price controls on most drug sales. Advocates claim that "reimportation" will allow American patients to benefit from this price disparity.

But "reimportation" doesn't actually describe what will happen if foreign drug importation is legalized. Using the term is an act of linguistic misdirection—or outright chicanery, if you've got a cynical streak.

Importing drugs from Canada is exceedingly dangerous for a number of reasons.

For starters, many internet pharmacies based up North are stocked with drugs from the European Union. And while many wouldn't hesitate to take medicines purchased from countries such as France and Great Britain, there's plenty of risk involved.

The EU currently operates under a system of "parallel trade," which allows products to be freely imported between member countries. This means that any drugs exported from the UK to Canada could have originated in an EU country with significantly less rigorous safety regulations, such as Greece, Portugal, Latvia, or Malta.

Just last year, EU officials seized over 34 million fake pills in just two months. And in May, Irish drug enforcers confiscated over 1.7 million pounds of counterfeit and illegal drug packages. So if American customers start buying drugs over the internet from Canadian pharmacies, they could easily wind up with tainted medicines of unknown European origin. It's also important to note that drugs from anywhere in Europe aren't even legal for sale in Canada. So, when politicians say we can get "the same drugs" that Canadians get, they're just plain wrong.

Even more worrisome is outright fraud—many "Canadian" pharmacies are actually headquartered somewhere else.

A 2005 investigation by the Food and Drug Administration looked at 4,000 drug shipments coming into the United States. Almost half of them claimed to be from Canada. Of those, a full 85 percent were actually from countries such as India, Vanuatu, and Costa Rica.

As part of another investigation, FDA officials bought three popular drugs from two internet pharmacies claiming to be "located in, and operated out of, Canada." Both websites had Canadian flags on their websites. Yet neither the pharmacies nor the drugs were actually from Canada.

As an FDA official told Congress, "We determined there is no evidence that the dispensers of the drugs or the drugs themselves are Canadian. The registrants, technical contacts, and billing contacts for both websites have addresses in China. The reordering website for both purchases and its registrant, technical contact, and billing contact have addresses in Belize. The drugs were shipped from Texas, with a customer service and return address in Florida."

And in laboratory analysis, every pill failed basic purity and potency tests.

Right now, American patients who head online to buy drugs are motivated by the cut-rate prices they see on the web. Health insurers could help patients avoid this temptation by reducing co-pays for drug purchases, particularly for low-income patients. If drugs become more affordable in the States, patients won't feel the urge to look for a bargain abroad.

Dropping drug co-pays would also help patients stick to their prescribed treatment regimes. All too often, people skip a dose, don't get a refill, or stop taking their drugs prematurely in order to save money. In the long run, though, not adhering to a drug regimen leaves patients less healthy—and increases national medical expenses by an estimated $300 billion annually.

Calling foreign drug importation "re-importation" is a clever way to sell the idea to the American people. But the term simply doesn't fit with the facts. In reality, Americans would end up jeopardizing their health by purchasing unsafe drugs made in foreign countries.

Foreign Medications Demand Improved Quality Controls

Originally published May 28, 2014

In the past year, several major foreign pharmaceutical companies have been forced to recall thousands of medications sold in the United States. The problems with their products were myriad, including packaging defects, dosage errors, and bacterial contamination. In one especially egregious instance, epilepsy medications were found in a bottle of diabetes drugs, resulting in the recall of more than 2,000 bottles produced by a company based in India.

Given these dangerous errors, it's little wonder that American physicians are growing increasingly worried about the safety of prescription medications imported from foreign pharmaceutical companies. Compromised medications jeopardize the health of their patients.

We must strengthen America's "pharmacovigilance." The federal government should bulk up its drug import oversight apparatus and implement a tracking system that monitors the entire lifecycle of medications.

Right now, 40 percent of America's over-the-counter and generic drugs come from Indian suppliers. The World Health Organization has estimated that around 20 percent of the drugs manufactured in that country are counterfeit.

As the *New York Times* recently noted, at one hospital in Kashmir, fake drugs likely led to the deaths of hundreds of infants. In January, the Indian health ministry identified at least 32 medicines being sold in Indian stores that didn't meet basic safety and quality standards. Earlier this year, the FDA's commissioner, Margaret Hamburg, expressed her dismay about "recent lapses in quality at a handful of [Indian] pharmaceutical firms."

To protect against dangerous drug imports, the FDA has established the Office of Pharmaceutical Quality, which will more meticulously monitor the quality of brand-name, generic, and over-the-counter drugs.

Congress also enacted legislation in 2012 mandating that foreign pharmaceutical manufacturers be subjected to more intense scrutiny. This bill led to 160 FDA-conducted inspections at Indian drug manufacturers in 2013, an enormous increase over the number of site visits in previous years.

What American inspectors discovered was often profoundly disturbing.

For example, at one plant owned by the Indian pharmaceutical manufacturer Ranbaxy, the FDA found flies "too numerous to count." Ranbaxy has since been hit with half a billion dollars in fines and has plead guilty to felony charges.

These developments are all the more discouraging once situated within the broader context of the global economy. Pharmaceutical manufacturing is a $14 billion industry in India. When its firms are permitted to push subpar medications into the U.S. healthcare market, they often end up displacing American companies.

Such displacement can hurt our economy. Drug firms currently support 810,000 jobs throughout the United States. More than 20 percent of the sector's domestic sales are directed back into local research and development projects.

The American pharmaceutical industry has already proven its commitment to ensuring the quality and safety of the medicines it produces. It has vigorously pushed for the federal adoption of a track-and-trace system, which would monitor a drug's journey from manufacturer to packager to distributer to pharmacy and, ultimately, to patients.

Such a system would significantly strengthen the security of America's drug supply chain.

In contrast, India is simply shrugging off its gross failures in drug production. The Indian Medical Association's secretary-general recently said: "Our drugs are being sold in many countries and being accepted, so we have no issues. How do I know that Western drugs are better than our drugs?"

That casual attitude is unacceptable. American consumers are being put at direct risk by India's malfeasance. If Indian authorities won't take responsibility, the United States must step in and ensure that proper vigilance is in place throughout the entire lifecycle of a drug.

A global pharmaceutical market benefits consumers only when the appropriate safeguards are in place. The recent spate of recalls should prompt renewed dedication to pharmacovigilance. For the sake of both American drug companies and American patients, federal regulators must work harder to ensure that all drug imports meet basic standards for quality and safety.

The Delusion of Foreign Drug Importation

Originally published September 17, 2018

The Trump administration has just announced the formation of a working group at the Food and Drug Administration to study legalizing the importation of foreign-made drugs. Although the study is limited to a narrow set of circumstances in which imports might be allowed, it sets us on a path we shouldn't step on.

The goal is to lower costs and prevent price gouging by drugmakers who find themselves with monopoly power over a particular treatment—say, an out-of-patent medication for a rare disease with only a single generic manufacturer in the market. But proponents of imports are only starting with the narrow case. Their real goal is to allow foreign imports on a massive scale.

Unfortunately, their efforts are based on a mistaken premise. Legalizing drug importation would not generate substantial savings. Worse, it could expose millions of Americans to dangerous counterfeit drugs.

The FDA study, if conducted honestly, will end up demonstrating that the dangers here far outstrip any potential benefit.

Several states have tried importation schemes in the past. They've all failed. Illinois, for instance, spent 1 million in taxpayer dollars launching its "I-Save-Rx" program. After much fanfare, only 0.02 percent of the population took advantage of the program during its first year and a half. The state ended the program shortly thereafter.

A similar program in Minnesota filled under 140 prescriptions per month, while one in Boston had just a couple dozen takers. Vermont's attempt never even got off the ground.

If Canadian drugs were really as cheap as importation advocates claim, why have all these small-scale importation experiments failed? Why haven't patients flocked to them?

Because the price disparity isn't nearly as great as importation advocates claim. Yes, some high-end specialty drugs cost more in the States, but the vast majority of domestic drugs are low-priced generics. In fact, about 85 percent of all medicines consumed in the United States are actually less expensive than those in Canada or Europe.

Consider Lipitor. A 30-day supply of the popular cholesterol medication can cost more than $200. But the same supply of the generic version can cost as little as $4. Buy the same generic supply from CanadaPharmacyOnline.com, and it'll cost more than $14. That's a cost 250 percent higher.

Correspondingly, the savings from bringing Canadian drugs stateside are small to nonexistent. In fact, the Department of Health and Human Services estimates that legalizing importation for the entire nation would drop total American drug spending by less than 1 percent.

Patients may also be leery of foreign drugs for safety reasons—and rightly so. The production standards and quality controls are much weaker in foreign nations, even developed ones such as Canada.

Just ask Kenneth McCall, the president of the Maine Pharmacy Association. After Maine legalized the importation of foreign medicines, McCall ordered four prescriptions from an online Canadian pharmacy. All of the medicines were manufactured in India and all of them were fake. One even contained toxins.

Unfortunately, this incident was no fluke. Most drugs coming in from Canada aren't actually Canadian in origin. According to the FDA, 85 percent actually come from other, less developed, countries. If people think our prosperous and friendly northern neighbor is offering its seal of approval on these products, they should think again. Canadian authorities offer no such guarantees of authenticity and safety.

The world is awash in counterfeit drugs. In just one week last summer, international agents seized 25 million illicit and counterfeit drugs worldwide.

It's no wonder that not a single Secretary of Health and Human Services, of either party, has ever been willing to declare that imported drugs are safe.

From the standpoint of reducing costs, foreign drug importation is a bad idea. From a safety standpoint, it's even worse. The FDA working group has been asked how the import process can be made safe. Even under the narrow circumstances the FDA is considering, the honest answer is that it can't be.

Drug Importation Bill Would Expose Americans to Counterfeit Meds

Originally published January 23, 2019

Senators Chuck Grassley (R-IA) and Amy Klobuchar (D-MN) just introduced legislation to legalize the importation of prescription drugs. The "Safe and Affordable Drugs from Canada Act of 2019" would enable Americans to purchase cheap medicines from Canadian pharmacies for their own personal use. The senators believe the bill would reduce patients' spending at the pharmacy counter.

Lowering drug costs is a noble goal. But importation is not the answer. At best, the bill would yield little savings. At worst, it could endanger American lives by opening the floodgates to harmful counterfeit drugs.

Drug importation is hardly a novel idea. Politicians have floated various importation schemes for decades.

Several U.S. states and cities have even set up importation programs, which mostly imploded shortly after their inception. For instance, Boston's mayor set up a program in 2004 to give city employees access to imported Canadian drugs. But only a few dozen people ever signed up for the program, which was closed down in 2008.

There's a good reason such programs weren't successful. For starters, Canadian drugs aren't as cheap as they're chalked up to be.

Importation advocates have long exaggerated the disparity between Canadian and American drug prices. They often point to a handful of specialty drugs that cost more in America. In reality, most generic medicines cost the same or less in the United States than they do in Canada. And generics account for roughly 80 percent of prescriptions filled in the United States. One study from the nonpartisan Congressional Budget Office estimated that legalizing prescription drug importation would reduce U.S. pharmaceutical spending by only 1 percent.

The safety risks of importation far outweigh any meager savings.

The FDA can't regulate Canadian pharmacies, so there's no way to verify whether the drugs purportedly imported from the Great White North are safe. During a weeklong anticounterfeiting operation last year, Canadian officials inspected nearly 3,600 packages—and found that 87 percent contained counterfeit or unlicensed health products.

A striking number of "Canadian" drugs aren't actually from Canada. Canadian internet pharmacies regularly import drugs from less developed and less regulated countries, such as Turkey. Then they slap on their own labels and ship them elsewhere.

One FDA operation found 85 percent of "Canadian" drugs originated in 27 different countries. And more than a third of those drugs were potentially counterfeit.

Such concerns explain why Illinois ditched its importation program, iSaveRX, in 2009 after failing to adequately inspect foreign pharmacies. According to a state audit, "40 percent of the required inspections of the foreign entities claiming to be pharmacies were never completed, putting patients at risk" and patients were left with "no regulator to protect them."

Canadian regulators have warned Americans that importation could be risky. One official at Health Canada, which oversees that nation's pharmaceutical supply, said the regulator "does not assure that products being sold to U.S. citizens are safe, effective, and of high quality, and does not intend to do so in the future."

Senior U.S. officials have issued similar warnings. Over the past 18 years, in both Democrat and Republican administrations, every FDA commissioner and secretary of Health and Human Services has failed to certify that importation is safe.

If Senators Grassley and Klobuchar's bill passes, more counterfeit drugs could find their way into the United States. Congress should think twice before putting the health of American patients in jeopardy.

CHAPTER THREE
Part D Reform

Since 2006, Medicare Part D has helped ensure more than 43 million American seniors have access to affordable prescription drugs. The program was built to rely on market mechanisms. Under Part D, seniors choose from a wide variety of drug coverage plans. With so many insurers competing for seniors' business, they have a strong incentive to keep costs low and benefits generous. Seniors, moreover, are pleased with their Part D coverage. A poll sponsored by Medicare Today found that 90 percent of enrollees are satisfied. Still, many politicians are willing to compromise this highly successful program through poorly designed "sound bite" solutions that deliver transitory headlines but would surely result in reduced access to lifesaving medications for America's Medicare population. We must never allow populism and politics to trump the public health.

No Way VA Is a Better Way

Originally published April 18, 2006

It seems like such a good idea.

The government has gotten into the business of administering prescription-drug insurance plans through the Medicare Modernization Act of 2003. But it turns out the government already buys some prescription drugs … for the Veterans Administration (VA). Why not model the new program after the older one, reap the benefits of the combined market clout, and achieve savings through uniformity? That's what many lawmakers are asking.

Unfortunately, it doesn't work this way. Drug companies make price concessions to veterans they won't make to even the largest pharmacy benefit managers (PBMs) with which they negotiate.

Under rules set by Congress, to sell drugs to the VA, companies must offer each drug at a price that "represents the same discount off a drug's list price that the manufacturer offers its most-favored nonfederal customer under comparable terms and conditions." The medication must be offered "at a discount of at least 24 percent off [the] nonfederal average manufacturer price (NFAMP). An excess inflation rebate is also required, equal to the percentage by which the price increase for [the] drug has exceeded the consumer price index (CPI) in the prior period."

The manufacturer must make all of its drugs available through the Federal Service Schedule for any of its drugs to be eligible for reimbursement under the VA and Defense Department health systems, the Public Health Service (including the Indian Health Service), the Coast Guard, and the various state Medicaid programs.

Why do drug companies even bother? The VA represents only about 2 to 3 percent of their market—so they're not losing too much money on the deal. The VA also serves as a training ground for new doctors, and the sales allow drug companies to expose these new doctors to their medicines. The VA operates a closed system, which means there is little risk the drugs will be resold. And drugmakers want to maintain a good relationship with the government.

Obviously, none of these factors apply to the Medicare program. Even if they did, it wouldn't be a good deal to model Medicare after the VA.

For one thing, the VA offers veterans fewer drugs—about 93, compared to more than 130 for most PBMs. And the drugs it does offer are not exactly top of the line.

A study by Professor Frank Lichtenberg of Columbia University found that the majority of the VA formulary's drugs are more than eight years old and more than 40 percent are 16 years old or more. Just 19 percent of all prescription drugs approved by the FDA since 2000 are available to veterans; only 38 percent approved during the 1990s are.

In many cases, veterans must try a drug that comes close to the one they need and prove that drug doesn't work before they can get the drug they truly need. And in almost every case, the VA imposes an automatic one-year hold on new medicines while it "studies" the effects. These are drugs approved by the Food and Drug Administration—the most rigorous testing regimen on Earth—for use by all Americans.

Is this any way to treat our veterans? Is it a system we want to graft on to the whole of Medicare?

More importantly in the big picture, is this a way to save money? Is it cheaper to force the 41 million Americans who are eligible for the Medicare drug benefit to use medicines that are approximately what they need rather than specifically what they need in the name of saving a few bucks by buying in bulk?

Experts say these savings come at the expense of the patients we're trying to help. Forcing patients to accept lower-priced, less-effective drugs can lead to higher total drug spending because doctors then need to prescribe more drugs.

Beyond that, it could add to total program costs by leading to more hospitalization and physician visits related to using less-than-optimal medications.

We hear a lot of half-truths with regard to drug pricing. California Bay Area Democrats in Congress recently released a report that said that if Medicare drugs could be purchased through the VA system, the government could reap savings of up to 75 percent. But the study deliberately omitted plans that allow patients to pay a fixed low price for all the drugs they need. It also didn't mention that fewer drugs are available through the VA system or that drug companies simply cannot afford to make those drugs available at those prices through the Medicare plan.

In other words, it's as my grandmother always said: a half-truth is a whole lie.

Dorgan Had It Right the First Time

Originally published August 11, 2006

Sen. Byron Dorgan (R-ND) just introduced legislation to remove price competition from the new Medicare drug program. Specifically, his amendment would give the government the "authority to negotiate prices with manufacturers."

But guess what? Senator Dorgan was against it before he was for it.

In May 2000, he co-sponsored the following amendment to the Medicare reform legislation:

> NONINTERFERENCE. —In administering the prescription drug benefit program established under this part, the Secretary may not—(1) require a particular formulary or institute a price structure for benefits; (2) interfere in any way with negotiations between private entities and drug manufacturers, or wholesalers; or (3) otherwise interfere with the competitive nature of providing a prescription drug benefit through private entities.

Senator Dorgan had it right six years ago. The chief actuary of the Centers for Medicare and Medicaid Services (CMS) said the government "would be unlikely to achieve prescription-drug discounts of greater magnitude than those negotiated by Medicare prescription drug plans responding to competitive forces."

The nonpartisan Congressional Budget Office has also said flat out that the government "would not be able to negotiate prices that further reduce federal spending to a significant degree."

Six years later, the prescription-drug benefit is up and running. And after some hiccups, which are to be expected in the ramp-up of any federal program of this size, the noninterference clause Senator Dorgan championed is leading to big savings.

Robust competition is driving down the costs of Part D for both beneficiaries and taxpayers, with the majority of seniors satisfied with the benefit.

The numbers are eye-popping. According to the CMS, the net cost of the drug benefit to the federal government over the next decade will be 20 percent lower than government estimates just last year. That's a savings of $180 billion.

CMS attributes the vast majority of the reduction to lower-than-expected drug costs and higher-than-expected price breaks—savings that were negotiated by private plan managers under Medicare Part D. That is to say, rebates and discounts on drugs sold through Part D plans are even larger than originally projected.

Senator Dorgan would look absolutely prescient right now except for one thing—his recent amendment:

> REQUIREMENT TO NEGOTIATE PRICES WITH MANUFACTURERS. —In order to ensure that each Part D individual who is enrolled under a prescription drug plan or an MA-PD plan pays the lowest possible price for covered Part D drugs, the Secretary shall negotiate contracts with manufacturers of covered Part D drugs, consistent with the requirements of this part and in furtherance of the goals of providing quality health care and containing costs under this part.

Talk about a flip-flop.

Six years ago Senator Dorgan understood that keeping government out of price negotiations lowers prices. He knew that government interference would limit seniors' access to drugs by forcing them to fail on one drug before they can try a newer treatment. Government "negotiations" would discourage research and development by as much as a fifth and compromise the quality of care for present and future generations.

Senator Dorgan should remember that private pharmacy benefit managers, or PBMs, have decades of experience in managing drug programs. Medicare has none. These private companies face consumer pressure to keep costs down. Medicare does not.

These private companies are also free to conform their plans to patient needs. It's in their interest to allow access to the right drugs the first time for all patients. Medicare likely would have to limit access to drugs to keep costs within limits, hurting the quality of patient care for Medicare beneficiaries.

As any Econ 101 student knows, competition reduces costs. Government intervention, on the other hand, reduces opportunity. Federal interference was a bad idea when Sen. Dorgan smartly opposed it in 2000. And it's still a bad idea now that he supports it.

A Promise Worth Breaking

Originally published January 12, 2007

One hundred hours. That's not a lot of time—just two and a half weeks of work for the average American. But apparently, if Speaker Pelosi and the Democrats keep all their promises, it's plenty of time to ruin our healthcare.

This Friday, as part of Pelosi's agenda for the first 100 hours of the new Congress, Democratic lawmakers promise to pass legislation allowing Medicare to negotiate prices directly with pharmaceutical manufacturers. Their idea is to use the government's large market share—some 42 million beneficiaries—as a big chip to negotiate dramatically lower prices.

There's one problem with their plan, though. That bargaining chip isn't nearly as big as they think it is.

If Pelosi and the Democrats had pledged to spend just a few more of their first 100 hours learning a few basic facts about economics, they would realize the folly of thinking that government can do business better than private businesses themselves.

Indeed, once Pelosi gets to the negotiating table, she'll find out that the government's bargaining chip isn't as big as she hoped.

In any negotiation, the side that can walk away from the table with the least amount of pain has the upper hand.

Pharmaceutical manufacturers, who often have a patent giving them the sole right to market a drug, will have the upper hand because there are no competitors the government can go to. More often than not, they're the only game in town.

The government's negotiator, on the other hand, will have a relatively weak position. If he walks away from the table, he'll face a political firestorm from seniors wondering why Medicare can't get them the medicines they need.

This is exactly what has happened since the Veterans Affairs (VA) department started excluding newer drugs from its National Formulary in order to control costs. As Columbia University researcher Frank Lichtenberg has reported, "Only 38 percent of the drugs approved in the 1990s, and 19 percent of the drugs approved by the FDA since 2000, are on the VA National Formulary."

This is why "a third of VA seniors prefers to switch to Part D, but cannot because they would lose other VA benefits," according to Manhattan

Institute scholar Benjamin Zycher. If Pelosi has her way on Medicare Part D negotiations, those veterans might not have any hope for a refuge after all.

Perhaps recognizing this problem, the House Democrats' proposed legislation, the Medicare Prescription Drug Price Negotiation Act of 2007, states that nothing in the bill "shall be construed to authorize the Secretary to establish or require a particular formulary."

This means the government's negotiator would not be able to ban any medicines because of a manufacturer's refusal to come down on the price. In effect, he would be forbidden from "walking away" from the negotiating table—the biggest source of bargaining power.

But if the government's negotiating position is so weak, how is it that private drug plans can do any better at negotiating lower prices with drug companies?

Paradoxically, it's because private coverage plans are smaller, and there are more of them. This means that the drug companies can "price discriminate," or charge higher prices to one plan and lower prices to another, leaving it up to the consumer to choose the plan that gives them the drugs they need at the best price.

This is the genius of competitive markets: They allow individuals to decide what's best for them, rather than be forced into a one-size-fits-all government plan.

This isn't obscure economic theory, either. It's been confirmed by real-world results elsewhere in the government, in Medicaid's drug plan.

According to a 1990 law, pharmaceutical companies must offer their wares to Medicaid at the lowest price charged to any private plan. Predictably, after that law was passed, drug companies cut back on the discounts they offered to private plans. In a recent report, the National Center for Policy Analysis concluded that "given the resulting price increases for non-Medicaid buyers—including other government drug purchasers, such as the Veterans' Administration—the savings to society are not at all clear."

On the whole, it's good for politicians to keep their promises, but it would be much better for our health if the Democrats broke this one promise. Don't worry, Nancy, we won't hold it against you.

For Seniors, No Need to Mind the Gap

Originally published July 14, 2009

The White House announced on June 22, 2009 that America's leading pharmaceutical companies have volunteered to finance, single-handedly, the expansion of Medicare Part D. Their pledge will help more than 3 million seniors afford prescription drugs that currently aren't covered by Medicare.

The groundwork for this plan was laid last month when a group of six representatives from organized labor and the healthcare industry met with President Obama at the White House and promised to help cut healthcare expenditures by $2 trillion over 10 years.

Just weeks later, only one of the meeting's participants—the Pharmaceutical Research and Manufacturers of America—has made good on its promise, pledging $80 billion to the health reform effort.

Roughly $30 billion will be used to expand the Medicare Part D prescription drug benefit. According to the program's current rules, patients are covered for drug purchases below $2,700. After that, only spending that exceeds $6,100 gets subsidized. The gap in coverage, often called the "doughnut hole," has long been criticized as the program's biggest flaw.

But, under the new agreement, most drugs purchased in the doughnut hole will be half price. This is a boon for the one in four Part D enrollees who currently fall into the gap. The remaining $50 billion will lower America's healthcare costs, although the details have yet to be worked out.

The reaction from many in Washington, however, has been tepid. Eager to demonize the drug industry, many have claimed that this money is not nearly enough to make a significant difference in health reform efforts.

Senator Olympia Snowe was quick to point out that the $80 billion represented only "a modest percentage" of prescription drug spending in the United States. Only in Washington would $80 billion be considered "modest."

Considering that the development of just a single drug sometimes costs as much as $2 billion, an $80 billion commitment over the next decade is a huge sum of money. In fact, it's roughly equal to what Pfizer, the world's largest drug company, will spend on research and development over that time frame.

The truth is that even if this pledge doesn't single-handedly end the healthcare crisis, it sets a valuable precedent. Cooperative relationships between the healthcare industry and the federal government will be necessary if meaningful healthcare reform is going to get passed.

President Clinton learned this the hard way. His 1993 healthcare reform initiative was done in, at least partially, by a coalition of industry groups—from hospitals to insurance companies—that were allied against him.

Aside from greasing the legislative wheels, collaboration between industry and government could also improve the quality of healthcare policy. For one, bringing the private sector into the reform process might safeguard us from ill-conceived legislation that would decimate the medical technology sector, stunting innovation and overburdening the economy.

For instance, many in Congress want to reform Part D so that the government can "negotiate" its own drug prices with pharmaceutical companies.

Such heavy-handed government intervention into the price of drugs, however, would have disastrous effects on medical innovation. In fact, it could reduce the number of R&D projects drug companies undertake by as much as 60 percent, according to the National Bureau of Economic Research.

By doing their part to reduce healthcare costs, pharmaceutical companies are giving the government less reason to meddle in the industry.

The agreement reached between America's drug companies and the White House is an important step forward in broadening access to high-quality medicine without destroying the private sector institutions that the nation relies on for medical innovation.

Changes to Part D Won't Save Medicare

Originally published February 4, 2013

With the debt-ceiling showdown approaching, Republicans are insisting on spending cuts in exchange for raising the country's borrowing limit.

President Obama says the debt ceiling shouldn't be used as a negotiating tool. But it's inevitable that additional cuts will be seriously considered. And, unfortunately, lawmakers are likely to take aim at Medicare Part D, the highly successful prescription drug benefit.

This approach is misguided. Cuts to Medicare Part D would have devastating consequences for seniors and taxpayers.

Created in 2006, Medicare Part D has helped ensure that America's seniors have access to prescription drugs. Today, nearly 47 million Americans are eligible for Part D.

The program was built to rely on market mechanisms. Under Part D, seniors choose from a wide variety of coverage plans. With so many insurers competing for seniors' business, they have a strong incentive to keep costs low and benefits generous.

This fierce competition has kept expenses down for seniors and taxpayers. Overall, Part D has cost 45 percent less than original estimates predicted—practically unheard of for a government program.

Part D monthly premiums will again average just $30 in 2013, according to officials at the Department of Health and Human Services.

Part D rates are even more impressive when compared to the overall health insurance market. The typical premium for an employer-sponsored family health plan rose 4 percent from 2011 to 2012 and 9 percent the year before.

Still, many Democrats are willing to compromise this program. Their proposal would require drug companies to pay a so-called rebate to the government for every medication sold to a Medicare Part D participant who also qualifies for Medicaid, the government health program for the poor.

The Obama administration believes this plan will save billions. But it could actually drive up costs for a most Medicare beneficiaries.

Just look at what's already happening in our Medicaid program. Drug companies are required by law to sell their wares to Medicaid at below-market prices, and consequently have been forced to raise prices elsewhere. In other words, the proposed Part D rebate program would effectively levy a new tax on every senior who isn't eligible for Medicaid.

Indeed, two independent studies have estimated that the rebate plan will raise premiums for traditional Medicare enrollees by 20 to 50 percent.

Higher premiums are a big concern for most Part D beneficiaries. According to the Medicare Today poll, 84 percent of enrollees were worried that a Part D restructuring would increase their out-of-pocket drug costs, while 53 percent were afraid it would cause them to cut back or stop taking their medicines altogether.

Our government health programs are still in serious need of comprehensive reform. Unfortunately, the president continues to attack the very part of the system that is most financially sound. Changes to Medicare Part D will raise only minimal revenue—and at the expense of America's seniors.

The Road to Single-Payer Healthcare Is Paved with False Promises

Originally published April 8, 2013

Congressman John Conyers just introduced a bill that would create government-run healthcare in the United States. This same proposal for a "single-payer" system has been put forth in every Congress since 2003, and like all of those previous bills, is destined to die.

Almost simultaneously, Senator Richard Durbin (D-Ill.) and Representative Jan Schakowsky (D-Ill.) introduced legislation that would require the Department of Health and Human Services (HHS) secretary to negotiate Medicare Part D drug prices with pharmaceutical manufacturers.

Such brazen attempts at socialized medicine are doomed to fail. But it doesn't mean the threat of government-run healthcare isn't real. In fact, the current push to allow federal price controls in the Medicare drug benefit, Part D, is a first step toward a single-payer system.

While Medicare as a whole is a fiscal basket case—due to run out of money in 2024—Part D has been the very model of a well-functioning federal program since its implementation in 2006.

The Congressional Budget Office (CBO) found that, between 2004 and 2013, Part D will cost an extraordinary 45 percent below what was initially estimated. Premiums for the program, meanwhile, are roughly half of the government's original projections. It's no wonder that 96 percent of beneficiaries are pleased with the program.

This success is largely due to Part D's market-based structure. Beneficiaries are free to choose from private coverage plans, forcing insurers to compete to offer the best options. Through negotiations with drugmakers, private Part D insurance plans have had great success in keeping pharmaceutical prices down.

Of course, anywhere there are market principles at work creating value for consumers, there are liberals eager to meddle—and Part D is no exception. It's hard to see the bill to require the HHS secretary to negotiate Medicare Part D drug prices as anything but a federal power grab.

But it's shocking that Democrats want to take a program that provides affordable medicine for millions of seniors and "reform" it in a way that limits drug access without saving money or addressing any of the systemic problems afflicting Medicare.

Democrats will never succeed in creating a single-payer system by passing one, all-encompassing bill. Instead, liberals in Washington will have to take over the health sector bit by bit. The push to impose Part D price controls is the latest attempt to grab a little more power for the federal government. Those who support a healthcare system that benefits from choice and competition have a lot to be concerned about.

Medicare Part D: A Miracle for Elderly Heart Patients

Originally published July 1, 2013

Congestive heart failure is the most common reason America's elderly are admitted to the hospital. This agonizing condition restricts patient blood flow, leaving them exhausted, wheezing, and short of breath. About 400,000 Americans are diagnosed with a congestive heart condition every year. Half die within five years of diagnosis.

In recent years, the most effective tool against this disease has proven to be breakthrough pharmaceuticals. New drugs help limit the effects of the condition and improve life expectancy. And the single more effective means of getting seniors access to such medications is the Medicare Part D prescription drug program.

Heart drugs can be quite costly, with an average prescription clocking in at $100 or more per month. Part D subsidizes drug prices for its enrollees. Without such coverage, many seniors suffering from congestive heart disease simply wouldn't be able to afford their prescription medication. They'd be forced to go without treatment and their condition would worsen.

Part D empowers seniors to follow their prescribed drug regimen and effectively combat their condition.

But Part D's benefit aren't confined to improved health. Getting patients these drugs early in the course of their condition actually saves the health-care system over the long run by obviating the need for more expensive and invasive in-hospital procedures.

A new study published in the *American Journal of Managed Care* found that when congestive heart patients did a better job of sticking to their prescribed drug regimens—thanks largely to Part D—the resulting medical savings totaled nearly $2.6 billion.

Researchers predicted that even greater improvements in patients' pharmaceutical usage could save Medicare an additional $1.9 billion each year.

This new study is just one of many highlighting the overwhelming benefits of Medicare Part D.

The Congressional Budget Office recently found total Part D costs are 45 percent below original estimates and in each of the last three years, the CBO has cut its Part D spending estimates by $100 billion. Meanwhile, a

Harvard study found that during the first two years of Part D's implementation, beneficiaries with limited prior drug coverage saw an average $1,200 reduction in their nonpharmaceutical health costs.

Medicare Part D's success derives from underlying market principles. Senior citizens can choose from plans offered by a wide variety of private insurance companies. Part D's "noninterference clause" prevents the government from meddling in the price negotiations between these private insurers and pharmaceutical companies. The competition drives down costs and improves services.

Not surprisingly, 90 percent of Part D enrollees are satisfied with their prescription drug coverage.

Nevertheless, the Obama administration and Democrats in Congress want to tamper with Medicare Part D's winning formula by implementing Medicaid-type "rebates." Their proposal would force pharmaceutical companies to cover a preset part of medications sold to Medicare beneficiaries who qualify as low-income.

It's essentially a way for the federal government to set price controls, intruding in the market and forcing the pharmaceutical industry to sell its products at prices below their real market value.

This would quickly weaken incentives for innovation in an industry where creating new drugs is already a risky and expensive process. Each new medicine costs around $1.2 billion and takes a decade or more to fully develop. Investors have no guarantee these efforts will ever pay off, and some might decide to cut back on research into new cures and treatments.

The rebate scheme may also raise costs for other beneficiaries whose income makes them ineligible for both Medicare and Medicaid coverage. For these 18.7 million people, monthly premium costs could increase by up to 50 percent a month, according to a study from the Lewin Group.

Medicare Part D has demonstrated its considerable merits, improving health, increasing access to much-needed medicines, and—most remarkably—cutting overall Medicare costs. Changes to Medicare Part D will undermine one of the most effective, market-based programs run by the federal government, hurting America's seniors in the process.

Stop Meddling With Medicare

Originally published April 14, 2014

As many as 36 million Americans narrowly dodged a recent proposal from the Centers for Medicare and Medicaid Services (CMS) that would have advanced unprecedented government intrusion within the Medicare Part D prescription drug program.

CMS never offered a satisfactory reason for its proposed tinkering with something that's working. Part D is a real rarity—a genuinely popular and cost-effective federal health program. The proposed reforms were not only unnecessary; they would have compromised affordability, limited treatment options for patients, and stifled innovation in the insurance industry.

Though the misguided proposal has been shelved, Part D is still at risk. The Obama administration has pledged to push through a similar plan in the near future.

By all accounts, Medicare Part D is overwhelmingly successful. The program offers drug coverage to seniors for a monthly average of $31 and enjoys a 90 percent approval rating among beneficiaries.

Unlike other healthcare benefits, Part D is also cost-effective for taxpayers. The nonpartisan Congressional Budget Office (CBO) reports expenditures are now 45 percent below original cost projections.

And because Part D provides access to vital medicines, the program helps seniors avoid expensive stays in hospitals and nursing homes. Over half of beneficiaries report they would be "more likely to cut back or stop taking medicine altogether" without Part D coverage. In this way, Part D saves our healthcare system $12 billion annually, according to a study in the *Journal of the American Medical Association.*

In light of the program's remarkable track record, it's inexplicable why CMS would attempt to tamper with important aspects of Part D.

As part of the proposed changes to Part D, CMS wanted to eliminate rules that provide "protected class" status to mental health drugs and drugs used for autoimmune diseases and organ transplant patients.

If the reform had taken effect, private plan providers may no longer have covered many of the medicines included in these designated categories. Countless Part D beneficiaries would have lost access to the drugs they were currently receiving and need to stay healthy.

CMS claimed that the proposal would lower prices by giving private plan providers better negotiating leverage with pharmaceutical companies. But that rationale ignores the fact that costs for these drugs have already been falling for years. From 2006 to 2010, in fact, prices for protected-class drugs dropped by 2 percent, despite an explosion in overall healthcare costs.

The idea that federal regulators can intervene to help negotiate better prices is dubious at best. And CMS didn't offer any quantitative evidence that restricting vital medications would benefit either seniors or the federal government.

Indeed, Part D works specifically because it encourages open competition between private insurers who are eager to serve Medicare's huge prescription drug market.

Nevertheless, federal regulators sought to reinterpret certain parts of Part D's "non-interference clause"—the part of the law that forbids regulators from distorting the market by intervening in drug price negotiations. The proposed changes would have allowed the government access to all agreements struck by drug manufacturers, pharmacies, and insurers.

The new rules would have also limited the number of bids an insurance plan may offer in a region to two. By reducing the number of plans available, CMS would have weakened competition between insurers while also preventing plan providers from experimenting with new policies. Uncle Sam wants to reduce competition? What would the Federal Trade Commission have to say about that?

On top of all that, the agency suggested changes that would encourage nearly every employer who offers health insurance for retirees to drop prescription drug coverage.

There was one common thread running through these proposed changes to Part D: they all restricted choice and discouraged competition. The Obama administration, it seems, isn't content with a healthcare entitlement that relies on market forces—no matter how cost-effective or popular the program may be.

When the Obama team shelved its plan, CMS Administrator Marilyn Tavenner only promised to pull it only "at this time." In fact, she committed to "advancing some or all of the changes in these areas in future years." So Part D isn't yet safe. In the future, if regulators want to sabotage the most successful federal health program in the country, the least they could do is explain why.

President Trump Misunderstands What Government Drug Price Negotiations Entail

Originally published February 1, 2017

President Donald Trump recently pledged to let federal officials negotiate the prices of drugs covered under Medicare. He claims this will save taxpayers billions of dollars.

Nobody doubts that Trump and his team are shrewd negotiators. But the sorts of "negotiations" that Trump refers to have nothing in common with haggling over a real estate deal. Instead, the action that Trump has proposed—repealing the noninterference clause, originally drafted by Democratic Senators Ted Kennedy and Tom Daschle—would result in Medicare drug prices going up and patient choice going down.

This clause has been the key to Medicare's success. Between 2004 and 2013, the Medicare Part D prescription drug benefit program cost an extraordinary 45 percent less than initial estimates. Premiums for the program also are roughly half of the government's original projections.

These unprecedented results are largely due to Part D's market-based structure. Beneficiaries are free to choose from a slate of private drug coverage plans, forcing insurers to compete to offer the best options to American seniors. This year, seniors can choose from among 746 plans nationwide, with an average monthly premium of around $35.

Such great choice and low costs have led to widespread support for the program. In fact, nine out of ten seniors report satisfaction with their Part D coverage, according to a recent survey.

Through their own negotiations with drugmakers, private insurers that offer Part D plans have had great success in keeping pharmaceutical prices down. In fact, the Congressional Budget Office (CBO) observed that Part D plans have "secured rebates somewhat larger than the average rebates observed in commercial health plans." The noninterference clause prohibits government officials from intruding in these negotiations.

Doing away with the noninterference clause, on the other hand, "would have a negligible effect on federal spending." In a report from 2009, the CBO reiterated this view, explaining that such a reform would "have little, if any, effect on [drug] prices."

In fact, allowing the feds to negotiate drug prices under Part D likely would have a negative effect on the program. The CBO explains that to achieve any significant savings, the government would have to follow through on its threats of "not allowing [certain] drug[s] to be prescribed."

In other words, the government might drop some drugs from Medicare's coverage. Patients who need those drugs would then be forced to pay for them out-of-pocket, which would make medicines vastly more expensive for the seniors that Trump wants to help.

If patients couldn't afford the prescription, then they might switch to a less effective drug or stop taking the medicine altogether. Their health would suffer.

Unfortunately, this isn't a hypothetical consequence. Just look at what's happening with the Veterans Affairs (VA) formulary, which permits government interference. The VA covers barely 80 percent of the 200 most popular drugs in the country. Medicare, which doesn't allow for government meddling, covers 95 percent of these medicines.

Letting Medicare go the way of the VA would be devastating for seniors.

Senators Kennedy and Daschle knew what they were talking about. The president should pay close attention.

Doggett Bill Destroys Drug Innovation

Originally published April 23, 2019

Getting Medicare costs under control is a full-time job. It's a big, urgent public health problem. But a bad solution isn't better than none at all and some proposed cost-cutting measures could end up doing much more harm than good.

Take the legislation just introduced in the House and Senate by Representative Lloyd Doggett (D-TX) and Senator Sherrod Brown (D-OH), respectively. Their "Medicare Negotiation and Competitive Licensing Act" is designed to reduce costs by fundamentally altering the highly successful financing structure of Medicare Part D, the prescription drug benefit for seniors.

Under their proposal, the federal government would step in to negotiate prices directly with drugmakers. The idea is that the government, with its vast buying power, can get a better deal than private payers. It's a common sound bite.

Unfortunately, even the government's own bean counters don't think there's any savings to be realized this way. And the unintended consequences of the Doggett-Brown proposal are even worse: Limiting therapeutic choices for seniors and gutting incentives for investors and researchers to develop the next generation of groundbreaking cures.

Medicare Part D is very popular among American seniors, and rightly so.

Under the current structure of the program, private-sector insurers negotiate directly with drugmakers on the price of their products for inclusion in insurance plans. Insurers pass these discounts on in the form of lower premiums. Seniors then choose from a wide variety of private-sector insurance plans and can usually find the one that best meets their individual needs.

By law, the government currently has to stay out of the price negotiations between insurers and drugmakers. The first thing the Doggett-Brown legislation does is lift this restriction, known as the "non-interference clause."

But there's a problem here. First, the government is unlikely to be able to get better prices from drugmakers than insurers are currently obtaining. That's the conclusion of both the Congressional Budget Office and the Office of the Actuary of the Centers for Medicare and Medicaid Services. Insurance companies already have an incentive to drive hard bargains, and drugmakers go along within the constraints of maintaining the viability of their businesses.

And that's where the second element of the Doggett-Brown proposal comes in. Their legislation would enable government regulators to seize the patent rights of innovative drug companies and give them to generics manufacturers.

Doggett and Brown are calling this theft "competitive licensing." That's a clever play on "compulsory licensing," a mechanism that's sometimes used in developing nations when a public health emergency creates the urgent need for the mass production of a particular medicine. When a compulsory license is invoked, a national government can allow generic drug firms to help fill a medicine shortage if an innovative drug company can't produce enough to meet public needs.

The proposed legislation would allow the U.S. government to seize patent rights whenever officials at Health and Human Services determine that a drug costs too much. This is extortion, backed up by the threat of theft. Vito "The Godfather" Corleone would love this legislation. But it's an offer we must refuse.

Few Americans would argue against lower drug prices. But patent theft comes with a price of its own.

Circumventing intellectual property protections discourages innovation. It costs approximately $2.6 billion and takes up to 15 years to shepherd just one new drug through the research and development pipeline.

Companies need a chance to make a return on such massive investments. Patents, which give drug innovators sole ownership of a medicine for a set period of time, offer that assurance. No competitor can create, market, or profit from that medication until the patent expires. This period of exclusivity allows research drug companies to recoup their huge investments.

The system works. The United States is the world's leading pharmaceutical innovator. Between 2001 and 2010, according to one major study, U.S.-headquartered companies were responsible for 111 of the 194 "new chemical entities" produced worldwide—more, that is, than all other countries combined.

U.S. drug firms would no longer be able to afford this groundbreaking research under the terms of the Doggett-Brown bill. They would be at risk of having the government step in to demand they turn over their patented products—or just as bad, accept whatever reduced price the government deems acceptable to prevent that outcome.

In other words, patients may never benefit from the next breakthrough cancer therapy or the long-sought cure for Alzheimer's. That's why the Medicare Negotiation and Competitive Licensing Act is an offer Congress really must refuse.

CHAPTER FOUR
Price Controls

You've heard it many times: "Why can't we have the same drug prices like they do in Europe and Canada?" Now consider this: We can—if you want prices to go up and choices to come down. Understandably, this must seem confusing but, as Oscar Wilde reminds us, "The truth is rarely pure and never simple." Let me explain.

About 90 percent of the volume of medicines prescribed in the United States are generic drugs—and these safe and effective treatments are less expensive than in either Europe or Canada. That's right—90 percent of the medicines prescribed in the United States are less expensive than across the Atlantic or north of the border. The remaining 10 percent are generally those newer medicines that treat more serious chronic and life-threatening diseases, the products that regularly commute the death sentences of, for example, people with various forms of cancer. And yet there are many running for high public office who would reinstate the death penalty for these patients. Let's look at the record.

America Spends a Lot on Healthcare—But So Does Everybody Else

Originally published September 16, 2008

On both sides of the aisle, politicians repeatedly criticize the amount of money America spends on healthcare.

They've got a point. America's $2 trillion healthcare tab is quite steep. But America's not alone—every other developed nation, even those with universal healthcare systems, struggles with high healthcare costs.

Indeed, people in other healthcare systems often pay more than Americans do, once taxes are taken into account. Add in the high nonmonetary costs of rationed or denied care and waiting lists, and suddenly the vaunted European systems commonly touted as models for the United States don't seem like a good deal at all.

Let's dive into the numbers.

In America this year, a family of four with an employer-based PPO will face about $15,609 total in healthcare costs. Of this amount, the employer will pay on average $9,442, and the employee will contribute $3,492 in premiums and $2,675 for co-pays and other expenses. Employee premiums are about 6 percent of the median family's annual income—less than what that family spends on food.

In Canada, while the percentage of taxes used to provide healthcare varies, it is estimated that 22 percent of taxes collected went to the health system in 2004. Several provinces, including Quebec, Ontario, Alberta, and British Columbia, also charge additional premiums. Canadians may spend their own money to receive private treatment for procedures or drugs that are not covered by the government system.

Citizens of the United Kingdom pay 11 percent of each pound they make in weekly income between $198 and $1,326 for care through the state-run National Health Service, plus an additional 1 percent of income over $1,326 per week. That's nearly double what Americans pay.

The co-pay for drugs is low, but many drugs are not covered, often because they are not considered cost-effective enough to justify inclusion in the government's plan.

But what if you need one of those drugs? Well, you can kiss your NHS benefits good-bye. Anyone who uses his or her own money to buy drugs outside the NHS will find him- or herself shut out of the system.

In Germany, coverage from a public sickness fund currently can range significantly in cost, from around 12.2 to 16.7 percent of income, with the employee paying a bit under half. This coming fall, premiums are set to be standardized—and healthcare experts anticipate that they will be set around 15.5 percent. Private patients can generally expect to pay more than they would in the public system.

In France, employees contribute only 0.75 percent of their salaries toward medical care, but they also pay a 7.5 percent General Social Contribution, the majority of which is earmarked for the health system. This base coverage reimburses people for the bulk of costs for doctor visits and for a portion of the costs of medications. On top of the government coverage, almost all French residents have supplementary coverage from a *mutuelle*, which costs approximately 2.5 percent of salary.

When compared to the U.S., the fact is that the healthcare systems in Europe and Canada don't save citizens much at all.

Health reform is urgently needed in this country, and cost-cutting will be a critical component of any reform efforts. Despite its supporters' claims to the contrary, government control of the healthcare marketplace is anything but a ticket to a lower-cost healthcare paradise.

Pink Ribbon Reality

Originally published October 30, 2008

You've probably seen a lot of pink ribbons around lately. Maybe you're even wearing one. October is Breast Cancer Awareness Month, after all.

But which presidential candidate deserves to wear that pink ribbon?

In other words, which of the two supports healthcare policies that are most likely to help in the fight against breast cancer? It's not enough to give lip service to the cause. Committing resources to funding research will help. But a variety of health policy issues affect women struggling with the disease now and all those who will struggle with it in the future.

Breast cancer used to be a death sentence. You might have watched the gruesome scene in last year's HBO miniseries *John Adams*, in which the second president's daughter Abigail endures an anesthetic-free mastectomy. Her ordeal didn't even save her life.

But incredible breakthroughs, many made in just the last decade, are now improving and even saving the lives of cancer sufferers. We might ask which candidate supports a program that will get women suffering from breast cancer the best care as early as possible.

Perhaps the most important new drugs given to cancer patients are biologics. These medicines are not the relatively simple chemicals that make up conventional drugs. These complex medicines, often given through injection, are created through the genetic engineering of living material. These are treatments researchers refer to when they discuss "gene therapies."

These revolutionary medicines are extending and saving lives around the globe. Herceptin, approved by the FDA a decade ago, actually stops a certain type of breast cancer cell from growing. Avastin is another biologic, approved by the FDA just this year. It can extend the lives of late-stage breast cancer patients by several months by stopping the formation of blood vessels that tumors use to grow.

Right now, these life-extending drugs are available here in the U.S. and elsewhere. But that could change if our nation moves toward a government-run healthcare system, as some candidates have urged.

Look at the United Kingdom. The government decides exactly which health treatments it will provide through the National Institute for Health and

Clinical Excellence (NICE). That body does cost-benefit analyses to determine whether a treatment is worth providing.

You can't put a price on the life of a loved one. But that's just what NICE does.

Pricier treatments are often the first thing to fall prey to penny pinching in strained government budgets. Biologics are in this category. The complex medicines take years to research and develop, and then more years to test and get approved.

That's why NICE originally wouldn't let breast cancer patients in the early stages of their disease receive Herceptin. The bureaucracy only changed its guidelines after a public outcry.

It hasn't changed its mind on Avastin yet. The UK will not fund treatment with this drug, even though it's been shown to extend lives. In fact, the government-run service is so adamant that patients don't use this drug that you'll be penalized even if you buy it yourself.

The NHS informed mother of two Colette Mills that if she insisted on taking Avastin, though paying for it herself, she'd have to foot the entire bill for her treatment—even for care the government typically provides.

Such a nightmare could happen here. Two senators have already introduced a bill to create a similar agency here to cut healthcare costs.

That's not the only way lawmakers want to cut costs to the detriment of care. We must also ask which candidate understands what makes the best climate for research and development of lifesaving drugs.

Some think we can lower costs for biologics the same way we have for conventional drugs, through generics. They want to pass laws making it easier for companies to manufacture what are called "follow-on biologics."

This might sound like a good idea, but comparing biologics to conventional drugs is like comparing apples to oranges. Biologics aren't made using chemicals—they're made using living tissue. There's no way to make an exact replica of gene therapies.

So if these drugs are legalized, lawmakers must make sure they're safe. And lawmakers must protect the financial incentives needed to create these drugs. Otherwise, we'll never know what lifesaving treatments we've lost.

So don't just look at each candidate's lapel this month to see if they're participating in the fight against breast cancer—look at their broader healthcare plans. The fight against breast cancer will take place on many different fronts.

From Eugene to Eugenics: Oregon's New Cost-Cutting Strategy: Deny Care to Cancer Patients

Originally published November 4, 2013

In an ad for the Affordable Care Act, local folk singer Laura Gibson plucks her guitar and muses about how the "Oregon way" is to "care for each one, every daughter and son." Though a little corny, the song delivers a worthy message: Oregon is committed to providing everyone access to healthcare.

The commercial is part of Oregon's new ad campaign to boost enrollment in the newly created "Cover Oregon" health insurance exchange. But the reality strums a different, far less organic tune.

Even as Oregon drops $3.2 million to spread their "each one" message, the Beaver State is also taking measures that would deny lifesaving treatments to desperately sick citizens.

In August, Oregon's Health Evidence Review Commission issued an update to its guidelines for providing cancer treatment to low-income individuals covered by the state Medicaid program.

These new guidelines require that Medicaid deny coverage for certain cancer treatments for patients who have been deemed "too" sick, haven't responded well to previous treatments, or can't care for themselves.

Through these new rules, Oregon state bureaucrats are severely restricting access to care and dooming potentially thousands of local patients to a premature death.

What's worse is that these new Medicaid guidelines are not grounded in the medical literature or best clinical practices, according to Kenneth Thorpe, chairman of the Partnership to Fight Chronic Disease. Rather, according to Thorpe, they're based "on the odds of survival observed in a group of patients."

It's true that for some late-stage cancer patients, the odds are long that any additional treatment can help. But without access to the latest that medical science has to offer, a patient's survival rate simply drops to zero.

As B.J. Cavnor of the Northwest Patient Education Network powerfully puts it: "Cutting patients off from cures means patients who could have beaten their illness will no longer have that chance." It's a frightening move from Eugene to eugenics.

These guidelines dictate that Medicaid provide only "palliative" care—painkillers, acupuncture treatments, wheelchairs, drugs for nausea, and the like. So while Oregon won't let Medicaid patients have access to cancer medicines that could prolong or save their lives, it will pay to make their deaths slightly less painful.

Is that what Oregon considers compassionate care?

About 19,000 Oregonians are diagnosed with cancer each year. Over 640,000 state residents are covered by Medicaid—that's about one in five of the total state population. And the state Medicaid ranks will swell next year, when the Affordable Care Act will raise the program's income threshold up to 138 percent of the federal poverty line.

Oregon's new Medicaid guidelines take treatment decisions out of the hands of doctors and patients and put them in the hands of distant state bureaucrats willing to cut costs no matter the human toll. It's the practice of cost-centric controls over patient-centric care.

Even supporters of the president's healthcare law have taken to calling these treatment restrictions a death knell for poor cancer patients. Cavnor, the patient advocate, has described them as "extremely frustrating and morbidly ironic, especially for those of us who have tried to argue that the Affordable Care Act doesn't allow for 'death panels'."

Is this really change we can believe in?

Promising to expand access to healthcare to all while denying it to those who need it most is brazen hypocrisy. Oregon should expect more from itself.

At the end of her song boosting the state's health exchange, Laura Gibson sings "live long, Oregon." That's a good aspiration. Oregon's state bureaucrats should live up to it.

Cheap Drugs "Like in Europe?" Not So Fast.

Originally published June 3, 2015

As John Adams said, "Facts are pesky things"—especially when they don't reinforce your beliefs. And nowhere is that more of an issue than in the debate over drug pricing.

A new study from the Institute for Health Policy compares drug prices in four countries and points out that countries such as Germany, the United Kingdom, and Australia have governmental regulations that help keep drug prices at "sustainable levels." The study authors urge policymakers to "do something" to change the situation in the U.S.

The new paper is full of statistics. But, as the saying goes, statistics are like bikinis—what they show you is interesting, but what they conceal is essential. One thing the paper conceals is that the majority of countries in the Organization for Economic Cooperation and Development spend more on pharmaceuticals (as a percentage of healthcare expenditure) than we do in the U.S., 15.9 percent versus 12 percent. That ranks us 26th out of 34 OECD nations in pharmaceutical spending.

In Europe, Australia, and Canada, brand-name pharmaceuticals tend to be significantly cheaper than they are in the U.S., largely because foreign governments impose stringent price controls on most drug sales. (The study conveniently neglects to mention that generic drugs are more expensive in Europe than in the U.S. And generic drugs represent 87 percent of all the drugs sold here at home.)

When it comes to new, innovative medicines for many serious and life-threatening diseases, government-dictated lower prices come with a very high price tag. Free healthcare like in Europe and Canada? Let's look at the record. Government-controlled healthcare is not free. It comes at great cost through higher taxes, wait times, and denials of coverage. According to the OECD, the French pay about 20 percent more in income tax while Canadians, according to the Fraser Institute, wait an average of almost 18 weeks from a general practitioner's referral to treatment by a specialist. In Canada about 22 percent of taxes go to the health system, and several provinces charge additional premiums.

Citizens in the UK pay 11 percent of each pound they make in weekly income, between £100 and £670 for the National Health Service (NHS).

And then there's an additional 1 percent of income over £670 a week. Although the copay for drugs is low, many drugs are not covered because they're not considered cost-efficient.

What can't be overlooked (and is conveniently absent from the Institute for Health Policy study) is that price controls equal choice controls. And choices (or the lack thereof) have consequences.

Consider the facts:

The five-year survival rate for early-diagnosed breast cancer patients in England is just 78 percent, compared to 98 percent in the U.S.

A typical Canadian seeking surgical or other therapeutic treatment had to wait 18.3 weeks in 2007, an all-time high, according to the Fraser Institute.

The average wait time for bypass surgery in New York is 17 days, compared to 72 days in the Netherlands and 59 days in Sweden.

More than half of Canadian adults (56 percent) sought routine or ongoing care in 2005. Of these, one in six said they have trouble getting routine care. Eighty-five percent of doctors in Canada agree private insurance for health services already covered under Medicare would result in shorter wait times. Approximately 875,000 Canadians are on waiting lists for medical treatment.

Now consider the Affordable Care Act. Patients can access any medicine they need—as long as it's on the exchange formulary. Sure, the ACA limits the degree to which insurers can charge higher premiums for sicker patients, but ObamaCare plans found a way around these rules: impose higher out-of-pocket costs for all or most specialty drugs. High co-pays effectively remove choice from the system for many patients.

The breakdown of Silver plans (the most popular category) is particularly revealing. In seven classes of drugs for conditions from cancer to bipolar disorder, more than a fifth of these plans require patients to shoulder 40 percent of the medicine's cost. And 60 percent of Silver plans place all drugs for illnesses such as multiple sclerosis and rheumatoid arthritis in the "formulary tier" with the highest level of cost-sharing.

Nearly every Silver plan across the country, in fact, puts at least one class of drug exclusively in the top cost-sharing tier. In effect, this leaves patients with a given condition—whether HIV or Crohn's disease—without a single affordable treatment option. Silver is the new black.

If we're going to look to other healthcare models for solutions, we must uncover and study their problems. Healthcare is too important to allow reform by sound bite. "Drugs from Canada" is as much a false promise as "free healthcare."

The Institute for Health Policy calls for policymakers to "do something"— and amen to that. What that something should be, however, isn't to copy foreign schemes that trade transitory short-term cost savings for long-term patient care. It's to support the continued innovation that saves lives and money—big money.

As Harvard University health economist (and healthcare advisor to President Obama) David Cutler has noted, "The average person aged 45 will live three years longer than he used to solely because medical care for cardiovascular disease has improved. Virtually every study of medical innovation suggests that changes in the nature of medical care over time are clearly worth the cost."

"The challenge," according to the PwC Health Research Institute, "may lie in targeting the patient most in need of the more expensive course of therapy."

Drugs aren't the cause of rising healthcare costs; they're the solution. Demonizing new treatments distracts from the real problem in U.S. healthcare: top-down cost-centric policies that focus on the near term, short-changing long-term patient outcomes, endangering "sustainable innovation" by denying fair reimbursement for high-risk investment in research and development. (Research and development costs big even if a drug never makes it to market—and most don't.) New treatments are a bargain. Disease is always much more costly. If we don't reward risk-taking on behalf of human health, both will suffer.

Referring to the Model T, Henry Ford famously said, "Any customer can have a car painted any color that he wants so long as it is black." That worked out fine—until there was competition. Choice is the great emancipator. The same is true when it comes to healthcare, but it's a lot more important.

Price Controls on Drugs: A Death Sentence for Innovation

Originally published August 25, 2015

Pharmaceutical pricing is a hot-button issue—even more so now that we're into a presidential election cycle.

Over a hundred cancer doctors recently criticized the six-figure price tags of certain cancer drugs. The nonprofit Public Citizen has slammed pharmaceutical firms for "needlessly high" medicine prices, specifically citing the $1,000-per-pill hepatitis C drug Sovaldi. Both groups have called for the government to step in and control prices in the drug industry. Bad idea.

Price controls are the wrong response to drug costs. By more effectively combating disease, innovative new treatments actually lower healthcare spending in the long run. Top-down price controls would drive dollars away from drug development, depriving future generations of new therapies and cures. Lawmakers must avoid innovation-killing price controls.

High-tech pharmaceuticals prevent patients from developing more serious conditions that require expensive procedures, hospital stays, or doctors' appointments. Solid research has shown that every dollar spent on advanced drugs cuts overall health spending by more than $6.

Consider the much-maligned Sovaldi. A 12-week course of treatment costs $84,000. The drug is more expensive than past hepatitis C treatments—but it's also vastly more effective. Solvadi cures 90 percent of patients. A decade ago, the cure rate was less than 50 percent.

This medical breakthrough averts much more expensive treatments. About one in five hepatitis C patients ultimately suffer cirrhosis of the liver. For a comparatively cheap up-front expense, Sovaldi stops these patients from needing liver transplants, which can cost nearly $600,000.

The medicine also cures patients three times quicker than earlier treatments and doesn't cause the same severe side effects. That means patients are more likely to complete the regimen, rather than wasting money by dropping out of treatment midway through, as many hepatitis C patients used to do.

Also, is anyone really paying "$1000 per pill?" Certainly nobody with insurance. And for those without coverage there are generous programs supplied by the manufacturer. What rates have large payers negotiated? Large payers negotiated discounts of between 40 and 50 percent off of the list price, but these discounts were not passed on to the consumer.

Sovaldi and other similar hepatitis C medicines prevent needless healthcare costs and enable patients to lead healthier lives.

Unfortunately, future generations of patients might not be so lucky. If the government steps in and ratchets back the prices of expensive medications such as Sovaldi, pharmaceutical innovation will grind to a halt.

After all, drug development is an extremely lengthy, expensive undertaking. Potential therapies have to undergo a rigorous testing process. More than 88 percent of tested drugs fail to secure FDA approval. Counting these failures, the average new treatment costs $2.6 billion and takes over a decade to develop.

Drug firms are willing to make such huge investments because they can recoup their costs through sales and earn a return. A blockbuster drug selling at market prices makes drug research and development worthwhile.

Price controls would squash that logic. It would be hard or impossible for developers to break even on new research projects. Already, just one out of every five pharmaceuticals approved for sale ever generates enough revenue to cover its research costs.

Many firms would stop investing in new drugs. If the United States had implemented European Union–type price controls back in 1980, up to $293 billion worth of research would have been lost, according to a recent study. Patients would have lost access to dozens of lifesaving pharmaceuticals.

Price controls could mean the next generation of treatments might never come to fruition. If America's leaders truly value medical progress, they must encourage medical innovation by rejecting price controls.

Healthcare Reform Through... Xenophobia?

Originally published November 16, 2018

What exactly did the president propose?

Lowering the price that Medicare pays for the prescription drugs it purchases.

The president couched his remarks as part of an effort to battle "unfair foreign prices." But none of his proposals addressed any change in the pricing policies of any other country. The villain wasn't the pharmaceutical industry, but "unfair foreign nations who freeload off of American drug development." The president's comments, as they say, were "for the fans, not for the players."

(It's also interesting that the president's proposal for referencing a basket of European drug prices does not include most of the globe's richest economies.)

How will this be accomplished?

The president wants the Centers for Medicare and Medicaid Services (CMS) to negotiate directly with manufacturers to achieve price parity with a basket of reference countries. By 2025, the target price decrease would (assuming all things work out according to a yet-undeveloped plan) average about 30 percent.

How soon will this happen?

These new strategies would be rolled out post 2020 via a pilot program that will be gradually phased in (and has not yet been developed).

Initially, this will take place via Center for Medicare and Medicaid Innovation (CMMI) pilot programs. Specific CMMI pilots would each address a specific product. For the plan to roll out across the entire spectrum, it would require federal legislation to revoke the existing noninterference clause that prohibits direct federal negotiations. This is why the president spoke about his desire for "bipartisan support." Political response from Democratic leaders has been tepid.

The noninterference clause was written by former Senators Ted Kennedy and Tom Daschle. Nobody doubts that President Donald Trump and his team are shrewd negotiators. But the sorts of "negotiations" that Trump refers to have nothing in common with haggling over a real estate deal. Instead, the action that Trump has proposed—repealing the noninterference clause—would result in Medicare drug prices going up and patient choice going down.

Through their own negotiations with drugmakers, private insurers that offer Part D plans have had great success in keeping pharmaceutical prices down. In fact, the Congressional Budget Office (CBO) observed that Part D plans have "secured rebates somewhat larger than the average rebates observed in commercial health plans." The noninterference clause prohibits government officials from intruding in these negotiations.

Doing away with the noninterference clause, on the other hand, "would have a negligible effect on federal spending." In a report from 2009, the CBO reiterated this view, explaining that such a reform would "have little, if any, effect on [drug] prices."

In fact, allowing the feds to negotiate drug prices under Part D likely would have a negative effect on the program. The CBO explains that to achieve any significant savings, the government would have to follow through on its threats of "not allowing [certain] drug[s] to be prescribed."

In other words, the government might drop some drugs from Medicare's coverage. Patients who need those drugs would then be forced to pay for them out of pocket, which would make medicines vastly more expensive for the seniors the president wants to help.

This clause has been the key to Medicare's success. Between 2004 and 2013, the Medicare Part D prescription drug benefit program cost an extraordinary 45 percent less than initial estimates. Premiums for the program also are roughly half of the government's original projections. These unprecedented results are largely due to Part D's market-based structure. Beneficiaries are free to choose from a slate of private drug coverage plans, forcing insurers to compete to offer the best options to American seniors.

What about Part B drugs?

Per Part B, the president announced a move from physician "buy and bill" payments based on the price of a product to a flat fee-for-service platform. This was previously tried during the Obama administration as a CMMI pilot program and worked pretty well. It would, among other things, remove the incentive for physicians to prescribe a more expensive product when less expensive options are available.

This will be particularly important for biosimilars. Such a change would require changes in both the strategies and tactics manufacturers use to incentivize physician prescribing. A "target price" will be set for Part B drugs, but even when the pilot program goes live, it will be slowly phased in (and we don't get to the new target price of drugs until 2025 at the earliest).

Per the Health and Human Services plan, initial CMMI "negotiation" pilots will focus on "single source drugs and biologicals, as they encompass a high percentage of Part B drug spending and are frequently used by physicians that bill under Medicare Part B." The focus of Part B drugs "in the line of fire" are largely oncology-related medicines that represent the highest gross cost to CMS.

Republicans and even some Democrats excoriated the Obama administration when it tried to reduce how much the government paid for these drugs in 2016. It's unclear what will be different this time around. The president's ideas—which haven't even been officially proposed as rules yet—face an uphill battle in Washington.

What's next?

The president's proposal lays out "potential calculation steps" and asks for feedback via an Advance Notice of Proposed Rulemaking. Comments must be submitted by December 31.

The president's proposals do not address the "other 90 percent" of healthcare costs in the United States. Those that have nothing to do with pharmaceuticals. Are we really willing to risk investment for development of innovative medicines by slicing and dicing 10 percent of healthcare costs but ignore the middlemen, insurers, pharmacy benefit managers, hospitals and other entities that consume the lion's share of healthcare spending in the United States?

Bottom line, nothing is going to happen quickly, and nothing will happen comprehensively at least until 2020 at the earliest.

Stay tuned.

HHS Secretary Azar's Drug Price Plan Is a Gift to Global "Freeloaders"

Originally published December 18, 2018

Prescription drugs are costly, especially in the United States. That's why, in an effort to lower the prices Americans pay at the pharmacy, President Trump recently pledged to end "the global freeloading that forces American consumers to subsidize lower prices in foreign countries."

Unfortunately, a new plan from the Department of Health and Human Services (HHS) undermines this goal.

HHS Secretary Alex Azar recently unveiled a new Medicare proposal that simply copies the foreign price-control policies that have enabled other countries to underpay for drugs for so long. Not only does the change fail to hold "freeloading" countries accountable—it will undermine access to life-changing medicines here in the United States.

Why are brand-name drugs cheaper in foreign countries than they are here in the United States? Simply put, it's because of price controls. In countries such as the United Kingdom, Canada, France, and Sweden, the government dictates what drug companies can charge for their products.

The prices of prescription drugs sold in Sweden, for instance, are determined by the Pharmaceutical Benefits Board. Canada's Patented Medicine Prices Review Board plays a similar role. So does France's Economic Committee on Health Care Products.

The single-payer healthcare systems in these countries are what make such price controls possible. When the government is the sole administrator of prescription drugs, it has extraordinary bargaining power—enabling it to secure below-market prices from drug companies.

In the end, American patients pick up the tab for these foreign discounts by shouldering much of the burden for global research and development. This is unfair—and something that President Trump is right to condemn.

Unfortunately, HHS's new Medicare proposal does little to rectify the situation. The policy would change what the government pays for drugs under Medicare Part B—a part of the program that covers medicines administered at a doctor's office, such as cancer therapies.

Under HHS's plan, Part B would base its drug payments not on a drug's average domestic price, as the program does now, but on what 16 other countries pay for the same medicine.

In no way does this reform hold countries accountable for their "freeloading" practices. To the contrary, it adopts foreign price controls right here in the United States.

Patients often lose access to the best medicines when their government adopts price controls. Of the drugs launched in the last seven years, only 60 percent were available in Sweden. And only half made it to patients in Canada and France.

Here in the United States, meanwhile, nearly 90 of those medicines were available. Americans will no longer enjoy generous access to the newest drugs if we embrace price controls.

There is a way to hold foreign countries accountable for their pricing policies—instead of just copying them. Doing so will require trade agreements that force these governments to value American pharmaceuticals fairly—and to protect the intellectual property (IP) behind these medicines.

The administration made tremendous progress in this direction with the recently forged United States-Mexico-Canada Agreement (USMCA). That trade pact holds our northern and southern neighbors to rigorous IP and pricing standards when it comes to sophisticated drugs known as biologics.

In the end, patients are the ones who benefit from agreements such as the USMCA, as these deals create an environment where medical innovation can flourish.

It costs an estimated $2.6 billion to create just a single new medicine, after all. That investment is only worthwhile if drug companies are confident that they'll have a chance to recoup their upfront costs. Trade pacts that ensure medicines will be fairly valued abroad provide that confidence.

Importing the socialist pricing tactics of foreign governments is no way to stand up for Medicare patients. To stop free-riding foreign governments in their tracks, the Trump administration should explore policies that ensure all patients pay a fair price at the pharmacy counter.

Embrace Biopharmaceutical Innovation to Save Lives

Originally published April 3, 2019

More Americans are beating cancer than ever before. Cancer mortality rates have plummeted 27 percent in the past quarter century, according to a new study from the American Cancer Society.

This encouraging trend is no accident. Drug companies have invented hundreds of breakthrough therapies in recent years, many of which stop cancer in its tracks. A majority of these medicines have been invented right here in the United States.

There's a reason for this. America's intellectual property (IP) laws are the best in the world. Drug developers flock to the United States because they know their products will be protected long enough to be worth the investment.

But now, some American politicians want to weaken IP laws, which they believe allow drug companies to set excessively high prices. Gutting IP protections would be a mistake—without them, research would come to a halt, and patients everywhere would suffer.

Drug research is enormously costly. On average, it takes more than a decade—and costs $2.6 billion—to bring just one new drug to market. These projects are also risky. Only a fraction of experimental drugs succeed. Just over 10 percent of medicines that enter into clinical trials are ultimately approved by the FDA.

America's strong IP protections encourage investors to fund medical research despite the high costs and risks. After all, no one would pour billions of dollars into projects with a high failure rate without the ability to profit.

Strong IP protections have made the United States the world leader in drug development. Investment in the U.S. biopharmaceutical industry hit an all-time high of more than $70 billion in 2017. Nearly 4,000 new medicines are currently being developed in the United States, including more than 1,100 that will treat cancer.

But not every country values IP. Around the world, countries pursue two major practices that threaten drug development.

Many developed nations arbitrarily cap what they're willing to pay companies for prescription drugs. In the United Kingdom, government bureaucrats unilaterally decide which drugs are worth spending money on. If drug companies won't meet their price point, they simply don't cover them.

Meanwhile, developing nations tend to disregard U.S. patents through a process known as compulsory licensing. In very specific circumstances—often in the case of a national health emergency—governments can issue compulsory licenses to domestic firms, which allow them to produce copies of patented medicines.

Unfortunately, many countries take advantage of the compulsory licensing process. Even without a public health crisis, countries such as India and Brazil allow firms to circumvent U.S. patents—just to save a few bucks on domestically produced knockoff drugs.

Each of these practices hinder U.S. firms' ability to recoup their massive R&D losses and hamstring medical innovation. The United States is often the only country where drugmakers can sell their drugs at a fair price.

But that may soon change, if certain politicians have their way. A recent Trump administration proposal would tie reimbursement for drugs covered by Medicare Part B to the prices paid in other developed countries, effectively importing price controls. Meanwhile, Representative Lloyd Doggett (D-TX) has introduced a bill that would empower the Department of Health and Human Services to grant compulsory licenses to American firms.

These policies would be as disastrous at home as they are abroad. If lawmakers really want to make drugs more affordable, they should take on foreign freeloaders—not imitate them.

Strengthening IP protections in trade agreements, such as the recently negotiated United States-Mexico-Canada-Agreement, is a great place to start. USMCA affords companies that produce biologics—drugs made from living organisms—a longer period of time when competing drug companies cannot access their clinical trial data. This ensures that innovators have ample time to profit from their discoveries before rival firms can produce their own version of a therapy.

Lawmakers should also work to end price controls abroad. If we axed price controls now, by 2030 we'd see an additional 13 drugs developed each year. Similarly, discouraging compulsory licensing schemes would enable U.S. drug firms to expand to new markets and earn more money for future projects.

Thanks to medical advancement, the average American lives a decade longer than they did in 1950. Even so, more than 600,000 Americans will die from cancer this year. Every 65 seconds, someone in the United States develops Alzheimer's disease.

The cure for any of these diseases could be just around the corner. Protecting intellectual property will help the patients of today get the medicines of tomorrow.

Price Controls Put Americans' Health at Stake

Originally published February 26, 2019

Scientists searching for cures to cancer, diabetes, Alzheimer's, and other deadly illnesses may soon lose their funding, due to a misguided new proposal from Secretary of Health and Human Services Alex Azar.

The proposal would gradually reduce Medicare's reimbursement rate for advanced drugs administered in hospitals, clinics, and doctor's offices by 30 percent. Secretary Azar claims these price controls "will save $17 billion in Medicare drug spending over the next five years."

It's true that price controls would save the government money, at least initially. But they'd also deter investors from pouring money into risky, expensive—but potentially game-changing—biopharmaceutical research projects.

The cuts to research funding would make it much harder for scientists to discover the cures of tomorrow. Those future medicines wouldn't just save lives—they'd also save the government money by stemming the rising tide of chronic disease. Secretary Azar's price-control proposal is penny-wise and pound-foolish.

Drug research is ludicrously expensive and fraught with pitfalls. Most research projects fail in the lab. And roughly nine in ten experimental drugs that emerge from the lab and enter human trials fail to gain FDA approval. This high failure rate explains why it takes almost $3 billion to develop just one drug.

Unlike most European nations, which impose strict price controls on medicines, the United States has a relatively free market for drugs. The ability to earn a sizeable return on successful drugs explains why a majority of world's new drugs are invented in America. America's researchers are currently developing more than 3,000 new medicines.

New treatments have transformed the way doctors and patients combat the rarest diseases. In 2017, researchers developed a therapy for sickle cell disease, a dangerous blood disorder, for the first time in two decades. They also discovered the first-ever treatment for Batten Disease, a neurological disorder.

Such treatments are saving lives. The United States has the world's highest five-year survival rates for most cancers, in large part because drug companies seek regulatory approval for their medicines in America's free-market system before seeking regulatory approval in price-controlled Europe. Fully 92 percent of all new medications are first launched in the United States.

Access to newer, more effective medicines accounts for nearly 75 percent of cancer patients' improved survival rates in recent decades.

Price controls would halt this medical progress. They'd make it nearly impossible for research companies to earn a return on their initial investments. As a result, the investors who currently fund drug research would redirect their capital to other business opportunities that offer better returns.

Consider how price controls have wrecked drug development in Europe. In the 1970s, more than 55 percent of all new drugs were developed in Europe. Just 31 percent were developed in America.

Now, those statistics have reversed, largely due to Europe's ever-stricter price controls, which have made America ever-more attractive for drug researchers. From 2001 to 2010, the United States generated more than half of all new medicines developed globally; Europe accounted for just one-third of all new drugs.

Price controls also force some countries to ration which drugs their citizens have access to. American patients have access to 90 percent of newly launched medicines right away. Patients in the United Kingdom, meanwhile, only have access to two-thirds. And Canadians can access just half, thanks to artificial price controls. We shouldn't aim to emulate systems that rob patients of effective cures.

Medical breakthroughs could save patients, and the government, billions of dollars by preventing or curing chronic disease. Approximately 1,500 innovative treatments targeting Alzheimer's, cancer, diabetes, heart disease, and stroke are currently in development. If researchers produced one successful treatment that delayed the onset of Alzheimer's by five years, the government could save over $200 billion annually by 2050.

Thanks to our relatively free market system, Americans enjoy unbridled access to breakthrough medicines. Our patients have some of the best health outcomes in the world precisely because we've avoided the pitfalls of socialist price controls.

Binding Arbitration Is Dangerous for Patients

Originally published April 22, 2019

Washington lawmakers believe they've found a bipartisan solution to lowering the cost of prescription drugs. A proposal called "binding arbitration" is being promoted by Speaker Pelosi's top advisors, who reportedly believe the Trump administration will be on board.

Let's hope the measure doesn't gain steam. Binding arbitration—essentially government price controls—would be catastrophic for patients. It would undercut medical innovation and bar American patients from accessing innovative treatments.

Binding arbitration was first floated by leftist health policy experts in 2008. More recently, a handful of Congressional Democrats, and even the Medicare Payment Advisory Commission (MedPAC), have voiced support for the scheme.

Binding arbitration would impact drug pricing negotiations between manufacturers and government-sponsored insurance plans, such as Medicare. Currently, these parties work directly to settle price points for prescription medications.

But sometimes the government and drugmakers don't agree on a price tag. And should these two parties reach an impasse, government officials could trigger the arbitration process.

It works like this: The government appoints a neutral, third-party arbitrator to settle the dispute. Drugmakers and the government would each make the case for their preferred prices to the arbitrator. And after considering arguments from both sides, the arbitrator would set the drug's price, which would be legally binding.

This might sound like a fair negotiation. But arbitrators are far from the neutral mediators they're made out to be.

For starters, arbitrators aren't accountable to the public. Indeed, they are unelected officials given exorbitant authority to set drug prices—all without having to own the consequences of their decisions if something goes wrong. This imbalance of power is worrisome.

Just as problematic, however, is that arbitration is really just a front for government price controls. The government has the power to appoint arbitrators without any input from drug companies. This means officials would almost certainly choose policy wonks who already agree with their drug pricing philosophies.

Put differently, the government always comes out on top. And instead of capping drug prices directly, the government merely appoints someone else to do their bidding. That's far from fair.

Price controls in any form pose a grave risk to patients. Consider that it takes an average $2.6 billion and 10 to 15 years to create a single new medicine. Investors undertake risky projects knowing they can recoup their upfront costs on the rare chance they successfully bring a new cure to market.

But price controls all but ensure that investors never profit. The government would undercut innovators by lowballing new medicines—no matter how much money or time went into creating them. So investors would hemorrhage large sums of cash in every research project. And as the model becomes unstainable, investors will wisely funnel their money elsewhere.

Less money flowing into pharmaceutical R&D means fewer new medicines for American patients. There are currently 4,000 drugs in development in the United States—that accounts for more than half of the drugs being researched and created worldwide. Unfortunately, patients may never benefit from these cures if the development pipeline dries up.

That's terrible news for Americans suffering from chronic diseases. Today, 60 percent of the United States' adult population has one chronic condition. Forty percent have two or more. That makes chronic disease the leading cause of death and disability in America—costing our health sector more than $3.3 trillion each year.

America can't afford for the government to stomp out medical innovation. But that's exactly what would happen with binding arbitration policies. For the sake of patients, let's hope lawmakers on both sides of the aisle abandon the proposal immediately.

CHAPTER FIVE
Fixing the System

The American healthcare landscape is an ecosystem, but it's all about one thing in the end—getting and keeping patients healthier. When it comes to new medicines, it's not just about research and development, FDA review, and manufacturing, it's also about whose fingers are in the drug pricing pie. That means the first thing we have to do is inject a large dose of transparency into the system. When people say, "my drugs are too expensive," what they generally mean is their co-pay at the pharmacy is too expensive. Why is that? Who sets co-pay costs anyway? The answer is crucial because higher co-pays are the leading cause of medication noncompliance. Let me introduce you to the mysterious middlemen of drug pricing, pharmacy benefit managers, (aka "PBMs"). If you haven't heard of them, it's because they've tried very hard to stay in the murky shadows—and for good reason. But the Era of Opaqueness for these "greedy intermediaries" is coming to an end. Why? Because smart reform should focus on ways to reduce what patients pay at the pharmacy counter—and sunshine is the best medicine.

A Prescription for Lower Health Costs... With No Co-Pay!

Originally published February 15, 2010

Alas, it's unlikely that Washington will pass meaningful healthcare reform any time soon. Yet healthcare costs are still exploding—making quality care unaffordable for too many Americans and putting a financial burden on us all.

Surprisingly, though, there's a smart move that health insurers can make that'll lower costs for consumers and insurers alike, and improve patient health: Reduce co-pays on prescription drugs.

High drug prices lead many Americans to skip doses or quit prescriptions entirely. Yet prescription drug prices aren't rising—it's patients' out-of-pocket costs, or co-pays. Over the past several years, insurance companies have become increasingly reluctant to foot the bill for brand-name medications.

Indeed, since 2000, co-pays have increased four times *faster* than prescription drug prices.

Patients respond to higher co-pays by skipping their meds more often. In 2003, researchers at Oregon University studied the effects of introducing a $2 to $3 co-pay for prescription meds among 17,000 patients. Adherence to treatment dropped by 17 percent.

Some insurers are even refusing to cover new prescription drugs. According to a study from Wolter Kluwer Health, insurers' denial rate for brand-name meds was 10.8 percent at the end of 2008—a 21 percent jump from the year before.

Abandoning treatment—a practice known as "nonadherence"—has serious consequences for patient health. For instance, people with hypertension who neglect their meds are over five times more likely to experience a poor clinical outcome than those who don't. Heart disease patients are 1.5 times more likely.

It also results in higher medical costs, as patients who go off their meds often end up in the hospital. Minor conditions that might have been controlled by inexpensive meds can sometimes balloon into life-threatening illnesses that require surgery or other costly treatments.

This makes sense. After all, a daily cholesterol-lowering drug is far less expensive than emergency heart surgery.

As Congress figures out what to do next on healthcare reform, private insurers can act now to control their own costs and vastly improve medical outcomes by lowering co-pays. State lawmakers and insurance regulators, too, should look to co-pay reform to help make healthcare more affordable for average Americans.

Transparency in Medicine Isn't a One-Way Street

Originally published April 10, 2016

A majority of Americans believe increased healthcare transparency should be a top national priority. It's easy to understand why. Rising healthcare costs, coupled with high-profile stories of price gouging at some small pharmaceutical companies, have left consumers feeling ripped off, especially when it comes to drug prices.

But most drug companies aren't whimsically increasing prices. In fact, if the healthcare industry were really transparent, people could see the truth: drug companies often aren't the culprits behind high costs. In fact, they're the best hope for dramatically lowering healthcare spending. The so-called pharmaceutical transparency bills under consideration around the country won't solve the price-gouging problem, but they will make it harder to create the medicines that will actually reduce healthcare costs.

The transparent truth is that the prices patients actually pay aren't set by drug manufacturers—they're determined by pharmacy benefit managers, insurers, hospitals, and pharmacies.

And these third parties frequently engage in … price gouging.

Consider the "prescription price shell game" uncovered in Minneapolis, where a local CVS jacked up the price of a kidney medication to more than $6 per pill from 87 cents. Or the Levine Cancer Institute in North Carolina, which collected nearly $4,500 for a colon cancer drug that hospitals typically buy for $60.,

Or consider New Jersey's Carepoint Health-Bayonne Hospital, which led a recent Health Affairs list of the nation's most egregious price-gouging hospitals. According to Health Affairs, Carepoint marks up patient care prices by more than 1,200 percent.

"They are marking up the prices because no one is telling them they can't," said study co-author Gerard Anderson.

Unfortunately, the media largely ignores such abuses, preferring to concentrate just on alleged misbehavior or greed by pharmaceutical companies. When one drugmaker released a breakthrough Hepatitis C cure, headline after headline blasted the company for the drug's initial $84,000 price tag.

Few follow-up stories have noted that, because of competition from other drugmakers, the manufacturer granted such big discounts—often in excess of 50 percent—that the medicine now costs less in the United States than in price-controlled Europe.

Even fewer stories put America's healthcare spending in perspective. Brand-name drugs accounted for just 7 percent of the $100 billion increase in healthcare spending from 2013 to 2014.

That 7 percent accounts for some of the most promising advances in treatment in decades. By addressing once-untreatable symptoms and complications, these advances help patients avoid expensive surgeries and lengthy hospital stays—which account for a far larger share of healthcare spending than pharmaceuticals do.

Journalists crying page one crocodile tears over high drug costs aren't just ignoring hospitals' and insurers' roles in jacking up retail prices. They're ignoring the fact that massive decreases in healthcare spending will only come about due to pharmaceutical cures. Better MRI machines are not going to end the scourge of cancer. New drugs could—and do.

Of course, medicines aren't cheap to create. The average cost of developing an FDA-approved prescription medication is $2.6 billion, according to the Tufts Center for the Study of Drug Development. That represents a 145 percent increase over the past decade.

For every successful new compound, hundreds of others once deemed promising end up abandoned. Research chemists at pharmaceutical companies may spend an entire career in the lab without working on a single drug that gets to market.

Understandably, pharmaceutical companies don't love to publicize their frequent failures. As a result, everyday Americans only see the successful, profitable drugs—and the high price tags that stem from the cost of research plus the markups tacked on by third parties.

Misguided activists in multiple states, including California and New York, are capitalizing on public anger about seemingly overpriced drugs to advance legislation that would require companies to disclose their profits on certain high-priced medicines, and the costs associated with developing them.

Such "transparency" bills won't paint a representative picture of pharmaceutical profits or stop healthcare price gouging, especially among hospitals, insurers, and pharmacies. But that's not the real purpose of the bills anyway.

The proposed laws are prerequisites to price control bills that would let the government cap drug prices. Transparency bills in Massachusetts and Pennsylvania already include such provisions.

Giving government the power to dictate prices would discourage the research investments that lead to better medicines.

Consumers are justifiably mad about healthcare costs. But their anger is misdirected. If the healthcare industry were truly transparent, Americans would see who's really to blame for rising prices. With rare exception, it's not the companies creating lifesaving medicines.

New Restrictions on Drug Coverage Are a Threat to Patients

Originally published November 9, 2016

Kareem Abdul-Jabbar's toughest fight wasn't on a basketball court.

In his early 60s, the six-time NBA champion was diagnosed with leukemia, the deadly blood cancer. Fortunately, Abdul-Jabbar had access to state-of-the-art medications, including the advanced drug Tasigna, which paralyzed his cancer cells and prevented further growth. Today, eight years after his initial diagnosis, Abdul-Jabbar is thriving and cancer-free.

Unfortunately, many of today's leukemia patients won't be so lucky. CVS Health, the nation's second largest pharmacy benefit manager (PBM) that oversees 65 million Americans' drug plans, recently rescinded coverage for Tasigna—and 130 other specialty drugs.

As a result, millions of people could be denied access to drugs that could save their lives. Instead of prescribing the medicines best suited to patient needs, physicians will be forced to recommend lower-quality treatments.

PBMs administer the prescription drug plans used by health insurers and employers. In recent years, these organizations have gotten stingy about which drugs they cover.

Back in 2012, the nation's largest PBM, Express Scripts, excluded no medicines from its list of covered drugs, while CVS Health left off about 30. Today, they exclude over 200, including an array of popular treatments for arthritis, Hepatitis C, and various skin conditions.

PBMs have also stopped paying for cutting-edge cancer treatments. In addition to Tasigna, CVS won't cover the revolutionary prostate cancer treatment Xtandi. Meanwhile, Express Scripts just stopped covering Zyclara, a cream that can help prevent skin cancer.

PBMs are restricting drug access in other, more devious ways as well.

CVS is also steering patients away from ultra-complex "biologic" drugs, forcing them to switch to lower-cost treatments the company claims are medically equivalent. But in many cases these less expensive therapies, known as "biosimilars," aren't approved by the FDA to be interchangeable with their brand-name alternatives.

Consider one study that compared the effectiveness of a Crohn's disease treatment and its biosimilar. An alarming eight in ten patients who took the biosimilar required a hospital readmission for additional treatment, compared to only one in twenty who took the original drug.

Despite these disturbing results, PBMs are comfortable forcing patients to use biosimilars and generic medications. That's because their only concern is bringing down short-term drug spending—even if those savings come at a cost to patients' well-being.

Ironically, this strategy will end up raising healthcare costs in the long run. If doctors can only prescribe less effective treatments, folks will get sicker, be hospitalized more frequently, and require more expensive care. That demand will drive up overall healthcare costs and overwhelm doctors and hospitals with waves of new patients.

That doesn't matter to PBMs, though. A dollar saved by avoiding top-notch drugs is a dollar that goes into PBMs' pockets—even if the patient becomes sicker on less effective treatments and racks up much larger hospital bills for insurers and patients to pay down the road.

PBMs coverage denials are a deadly prescription for America's patients. By shrinking coverage for cutting-edge treatments, PBMs are forcing sick people to use substandard drugs. It's about time patients mount a full-court press against this callous behavior.

Who's Stealing My Savings?

Originally published January 4, 2018

Payers have always hated co-pay cards because payers love restrictive formularies and hate patient choice. Witness UnitedHealthcare's new program to discourage patients from using innovator company co-pay coupons.

These coupons have been a thorn in the side of payers because they lead to the use of more-expensive drugs, higher utilization, and an increase in payer costs. Alas, they're a good deal for patients. But after all, aren't payer profits more important than either patient care or patient choice?

The writing's been on the wall for those paying attention. There's already a UnitedHealthcare program in place that calls for specialty pharmacies in its network to decline to redeem co-pay coupons for off-formulary products. The new arrangement expands on that pilot program by punishing patients who use co-pay cards—the majority of whom have high-deductible plans and those with coinsurance linked to out-of-pocket maximums.

Translation? If you went for a low-premium/high-deductible ObamaCare plan you're going to get it right in the neck. If you like your health insurance, you can keep it? The biggest lie in American healthcare has never been so topsy-turvy. In a healthcare ecosystem governed by the Affordable Care Act (AC), what happens when you don't like your health insurance and there isn't another game in town? What happens when your insurance provider decides to change the rules of the game to disadvantage the healthcare consumer in order to pad their own bottom line? Is that in keeping with the spirit of the ACA legislation?

Previously, both co-pay and the amount covered by the coupon were applied to the patient's deductible or out-of-pocket cap. UnitedHealthcare's change will likely cause deductibles and out-of-pocket limits to be met more slowly, which will drive consumers toward those products preferred by the payer rather than by patients and their physicians.

UnitedHealthcare's scheme isn't going unnoticed. According to a letter sent out by the California Rheumatology Alliance (CRA):

> It has come to our attention that a new 'Accumulator' program has started to roll out that will directly affect a patient's ability to use co-pay cards for biologic therapies dispensed through pharmacies. We need to brace and prepare for the reality that all four major national PBMs will seek to exploit these 'Accumulator' programs

starting January 1, 2018 … The program stops all payments from patient co-pay cards, which are used to pay for biologics and other specialty medications dispensed by pharmacies, from being applied towards deductibles and out-of-pocket maximums. Consequently, co-pay cards given to our patients will only pay out-of-pocket costs up to the card limit. Once the card funds have been exhausted, the patient will be hit with the shock of having to pay the deductible all over again.

In essence, both the insurance companies and pharmacy benefit managers (PBMs) are collecting the deductibles and out-of-pocket maximums twice with these "Accumulator" programs. According to best estimates, up to 20 percent of commercial medical insurance policies sold in 2018 will have these programs built in. The most alarming concern is that most employers who have purchased/are purchasing these plans are unaware these programs are present in the coverage. They have no idea how it will adversely affect their employees' care.

And, per the CRA, "We fear that access to the essential medications needed to manage our patients' conditions with be severely interrupted. This appears to serve only the purpose of increasing PBM profits while providing a small discount to medical premiums."

It's a sad story of PBM profits over patients. In the United States, nearly $15 of every $100 spent on brand-name drugs goes to PBMs, which claim they lower drug costs. However, the share of annual drug price increases that PBMs pocket—as opposed to pass on to consumers—has soared from 5 percent in 2011 to 62 percent in 2016. Three large PBMs control 78 percent of the market and use this market power to control what medicines people can use, what they pay, and where they get their prescriptions filled.

PBMs use clawbacks of retail pharmacy revenue, spread pricing on generic medications subject to maximum allowable cost pricing, and nonmedical drug switching to increase revenue. Clawbacks in the form of direct and indirect remuneration fees PBMs extract from pharmacies that push many Medicare consumers into the program's catastrophic coverage, driving up total spending.

While it may be true that co-pay assistance programs for some brand-name drugs with generic competition provides an "end run" around legitimate PBM cost-reduction strategies, it is completely untrue for most specialty drugs. The uniform elimination of co-pay assistance as a method of meeting "accumulator" requirements does not appropriately recognize this not-so-subtle difference.

Co-pay accumulator policies leave patients financially exposed and prone to noncompliance. But it's good for the PBM bottom line. Perhaps a better term for "co-pay accumulator" is "PBM bottom line enhancer." Sometimes it's worthwhile to call something by its proper name.

According to Susan Pilch, vice president of policy and regulatory affairs at the National Community Pharmacists Association, "PBMs are sitting in the cat bird seat in the pharmaceutical supply chain. They take advantage of the fact that they have unparalleled insight into everything that goes on upstream as well as downstream and have managed to build in revenue streams for that."

And if PBMs are the cat … who's the canary? As they like to say inside the Beltway, "If you're not at the table, you're on the menu."

Trump's Drug Plan Is a Winner

Originally published June 4, 2018

President Trump just rolled out his "American Patients First" strategy to bring down drug prices. The plan includes several promising reforms that could sharply reduce out-of-pocket costs.

Already, the administration has made progress in curbing the nation's drug spending.

After FDA Commissioner Scott Gottlieb won Senate confirmation last spring, he introduced a Drug Competition Action Plan to streamline the approval process for generic drugs. Last year alone, the FDA approved over 1,000 generics—an all-time record. Thanks to the newly available, cheaper treatments, Americans saved nearly $9 billion.,

Now, the administration is turning its attention toward pharmacy benefit managers (PBMs). These middlemen work on behalf of insurers, corporations, unions, municipalities, and other entities that sponsor health plans. PBMs act as the arbiter for all things drug-related in these plans. They determine which medicines are covered and how much patients will pay at the pharmacy.

PBMs use their gatekeeper role to demand large rebates from drug companies. If manufacturers don't acquiesce, PBMs can blacklist their products. Three PBMs—OptumRx, CVS Caremark, and Express Scripts—control over 80 percent of the market, so they have enormous leverage.

On average, PBMs receive rebates and discounts that reduce medicines' list prices by about one-third. But they only pass about 10 percent of these rebates to patients. They share the rest with their clients—or pocket the savings for themselves. And they keep these figures under wraps. So neither patients nor health plans know the full scope of the rebates.

Anthem, one of the nation's largest health insurers, actually sued its PBM in 2016 for allegedly overcharging by $3 billion each year and failing to pass on rebates. Patients have filed numerous class-action lawsuits against PBMs, alleging that the middlemen overcharged them.

President Trump wants PBMs to be more transparent about the rebates they secure. His administration may mandate that PBMs in the Medicare Part D program pass one-third of the rebates along to Medicare beneficiaries. This would save patients more than $19 billion over the coming decade.

He also floated another major reform—requiring PBMs to act as "fiduciaries" to their health plan clients. This would legally obligate PBMs to negotiate the best deal possible for their clients—rather than structuring rebates in a way that primarily benefits the PBM.

President Trump also wants to eliminate the pharmacist "gag rule"—a restriction imposed by PBMs and insurers.

Here's how it works. In many cases, a patient's co-pay exceeds the total cost of a drug. For instance, a 30-day generic prescription might cost only $10, yet the patient faces a $15 co-pay.

The rule prohibits pharmacists from informing patients that it'd be cheaper to just pay the $10 directly—without using their insurance cards. This forced silence costs American patients millions of dollars in unnecessary spending. Overpayments impacted one in four patients and totaled $135 million in 2013, according to a recent study from the University of Southern California.

Of course, the president's plan isn't perfect. It proposes price caps on Medicare Part B drugs, which include most chemotherapies and other intravenous medications. The proposal would require manufacturers to pay a rebate to Medicare when the price of their drugs increases faster than inflation.

Price controls deter researchers from developing new treatments. They trade short-term savings for patients' long-term health.

With a few exceptions, the American Patients First agenda is full of smart, practical reforms. The sooner it's implemented, the sooner people will enjoy lower prices at the pharmacy counter.

PBMs Are Hogging Patients' Drug Discounts. Kudos to Trump for Taking Them On

Originally published August 28, 2018

The Trump administration has aimed a dagger at the heart of high pharmacy bills: the go-betweens in the supply chain that have been gouging insurers, drugmakers, and, most importantly, consumers.

Through its blueprint to reduce drug spending—called American Patients First—the administration has invited public comment on the role and responsibilities of pharmacy benefit managers, or PBMs. These are the megacorporations that health insurers hire to administer their drug benefits.

Officials surely got an earful. In their role as go-betweens, PBMs negotiate big discounts from drugmakers. But rather than passing those savings on to insurers and consumers, they have found ways to pocket many of the proceeds for themselves.

The sums involved are huge. If just one-third of the total discounts PBMs negotiate for Medicare's drug benefit were passed on to patients at the pharmacy counter, seniors and others in Medicare would have nearly $20 billion more in their pockets over the next decade. Indeed, the top three PBMs—Express Scripts, CVS Caremark, and OptumRx, who together control 75 percent of the market—collected more than $10 billion in profits in 2015.

How are PBMs generating such profits?

A recently leaked contract reveals the sordid details of the tactics PBMs use to secure savings for themselves.

First, PBMs tend to hog rebates. Drug manufacturers offer PBMs a percentage discount or negotiated "rebate" off a drug's list price. In exchange, the PBM agrees to cover the drug under its affiliated insurance plans, thus making it available to more consumers.

The majority of rebate dollars should, in theory, flow back to health plans. Then the plans can pass those savings to patients in the form of lower premiums or out-of-pocket expenses.

But the leaked contract with Express Scripts, the nation's largest PBM, gets crafty with its definition of "rebate." The contract carves out so many exceptions to the "rebate" that the PBM itself ends up retaining most of the negotiated discounts for itself.

For instance, PBMs receive extra "inflation payments" from drug manufacturers to cover annual increases in a drug's list price. But these payments are not classified as rebates. Similarly, PBMs charge drug companies "administrative fees," "service fees," and more for their services. But these fees also aren't considered rebates. So PBMs don't pass them along to health plans. They keep them all for themselves.

Second, the document reveals that Express Scripts uses cryptic algorithms to reclassify generic drugs as brand-name drugs. Drugmakers usually offer higher discounts for generic drugs than they do for brand-name drugs. If a PBM gets a big discount on a generic drug, and then reclassifies a generic drug as a brand-name drug, it avoids passing along the full discount to the health plan it's representing. The PBM essentially buys the drug low, sells it high, and pockets the excess.

Third, PBMs underpay pharmacists. They do this by determining how much to pay pharmacies for prescription drugs and how much to charge health plans for the same drug, according to something called a Maximum Allowable Cost, or MAC. MACs, however, allow PBMs to charge health plans far more than what they pay pharmacies. It's the pharmacies—particularly independent, nonchain stores—that are hit especially hard by this practice. They wield little negotiating power to change specific terms of PBM contracts.

And if the pharmacy owners want to complain? The contracts may contain a silencing clause—as the leaked contract did—that prohibits parties from making derogatory statements about PBMs.

PBMs are knowingly profiting off the backs of American patients, pharmacists, and innovators. It's time for them to change their shady practices—and pass on discounts to patients. The sunlight the Trump administration is calling for will be a great disinfectant.

Healthcare Chutzpah vs. Health Plan Literacy

Originally published October 17, 2018

Co-pay accumulators—the newest trick that pharmacy benefit managers (PBMs) use to pad their pockets—harm both patients and the public health by making lifesaving medicines unaffordable. And to make matters worse, these programs are being sold to patients as a "benefit." The healthcare chutzpah of the PBM industry is astounding.

A co-pay accumulator is a PBM tool to ensure that the value of a co-pay card or coupon doesn't count toward a patient's out-of-pocket maximum expense.

In the United States, nearly three-quarters of all prescription claims are handled by the top three companies: Express Scripts, CVS Caremark, and OptumRx. Total profits for the three exceed $17 billion. A Health Affairs study estimates an additional $22.6 billion in gross profits for all of the remaining PBMs. Much of this profit is made on the backs of patients.

Over the last five years, according to the Department of Health and Human Services, pharmaceutical spending has increased by 38 percent, while the average individual health insurance premium has increased by 107 percent. Rebates, discounts, and fees haven't slowed precipitous premium increases. Co-pay accumulator programs will only accelerate this dangerous trend.

Until recently, manufacturer couponing programs limited or eliminated patients' out-of-pocket costs until their insurance deductible was met. These coupons are particularly crucial for patients suffering from multiple sclerosis, various cancers, rheumatoid arthritis, Crohn's disease, pulmonary arterial hypertension, psoriasis, and HIV/AIDS (among others). According to two new studies, co-pay accumulator programs (which don't allow a coupon's dollar amount to count toward reaching that deductible) significantly decrease patient adherence by creating a payment abyss known as a never-ending deductible (NED). NED talks. Patients suffer.

Per a study conducted by ConnectiveRx of 503 patients being treated for serious chronic conditions that frequently require specialty medicines, 75 percent of patients who used a co-pay coupon at least once in the last 12 months said co-pay accumulator programs would make it harder for them to adhere to their medicines. Not surprisingly, 66 percent believe such programs are unfair.

These patients pay 10 times more out of pocket than healthy patients and are forced to try cheaper or more rebate-rich drugs before getting medicines that work best. Faced with higher out-of-pocket costs and barriers to access, people are more likely to stop their treatment, getting sicker and more expensive to treat.

It's also important to note that PBMs do not share their rebate savings at the point of sale with patients while enforcing accumulator programs. The sharing of savings occurs in every other segment of healthcare (physician services, hospitalizations, dental services), but not pharmaceuticals.

Why is this important to all of us? A review in the *Annals of Internal Medicine* estimates that a lack of adherence causes nearly 125,000 deaths and 10 percent of hospitalizations and costs the already strained healthcare system between $100 billion and $289 billion a year. That's why.

The problem is growing larger. Many low-premium, high-deductible insurance plans are now incorporating co-pay accumulator programs—and, per Magda Rusinowski, vice president of healthcare cost and delivery at the National Business Group on Health, many employers are now only offering such plans to their employees. As such, educating employees about co-pay accumulator programs becomes essential. But what is the state of patient/employee knowledge of the NED issue—particularly urgent since we are now entering the annual open-enrollment period?

The ConnectiveRx study found the level of co-pay accumulator awareness at 25 percent. And, if knowledge is power, ignorance is not bliss. A new report from McKesson found that not only do patients not understand what a co-pay accumulator program is—what they believe to be true is wrong.

For example, 60 percent believe co-pay accumulators are a plan "benefit," and less than 40 percent know what "out-of-pocket" means.

This confusion is augmented and abetted by the misleading nomenclature used by health plans and PBMs. For example, three of the biggest PBMs promote their programs with terms such as "Coupon Adjustment: Benefit Plan Protection," "Specialty Copay Card Program," and "Out of Pocket Protection Program." The facts speak otherwise.

According to the McKesson report, "Patients can think these protection programs are a health plan benefit or a medication support tool." The reality is that these programs are being marketed to employees (via materials created by the PBMs) as cost-saving tools that make patients pay a "fair share" for specialty medicines. The net effects are that targeted patients (those with "expensive conditions") pay more for their healthcare.

One way to even the playing field is to recognize and address the need for patient programs designed to enhance "health plan literacy." Considering that 72 percent of healthcare-related bankruptcies impact patients with health insurance, this should be a public health priority. "Counter detailing" against well-funded and highly motivated PBM marketing schemes will be difficult and expensive.

Who should create and run these programs? It will have to be a collaborative effort between patient organizations (whose members are being impacted), physicians (whose patients aren't getting the medicines they need), and employers (whose employees are being misdirected). Eighty percent of plan sponsors (aka "employers") haven't looked at their PBM contracts in 10 years. Employers must step up to the plate and do the right thing for their most valuable resource—their employees. It will take time, effort, and ... integrity.

Knowledge is power. Health plan literacy begins with honesty and sunshine, the best medicine. This is precisely why PBMs loathe public attention. Too bad—but not too late.

Lawmakers Should Cap Patient Cost Sharing in Medicare Part D

Originally published June 6, 2019

The U.S. House of Representative health committee leaders have unveiled a discussion draft of a new bill designed to cap a Medicare patient's out-of-pocket spending in the Part D prescription drug benefit.

America's sickest seniors should be delighted by the news. Capping patient out-of-pocket spending will ensure they can access the drugs they need to live healthy lives.

Congress should see this effort across the finish line—and the sooner the better.

Medicare Part D has been the poster child of popular, successful, and budget-friendly government entitlements since its launch in 2006. Today, roughly 43 million seniors rely on Medicare Part D to access necessary medications.

Beneficiaries are largely pleased with their coverage; fully eight in ten seniors say their drug plan is "a good value." In short, when it comes to prescription drug coverage, Part D gets an A.

Part D's unique structure is one reason for this success. The government subsidizes coverage, but private insurers administer each plan. This forces insurers to compete with one another to create high-value, low-cost options to earn customers' business.

And it's this competition that keeps prices at bay—for beneficiaries and the government. The average patient premium in 2019 is just $33.19. And Part D's program costs were 45 percent lower than predicted after a decade. That's largely unheard of when it comes to government entitlements.

But for some, the program isn't all sunshine and rainbows. The sickest beneficiaries—many of whom take multiple drugs—spend a small fortune on their medicines. That's because most plans require them to pay a certain co-insurance or co-pay fee for each prescription they pick up.

Consider that in 2016, the latest year for which data is available, more than 800,000 Part D enrollees spent $5,000 or more at the pharmacy counter.

That's a lot of money—especially when you consider that more than half of Americans don't have enough money in the bank to cover an unexpected $1,000 emergency expense.

High out-of-pocket costs aren't simply a financial concern; they're also a threat to patient health. Studies show that high out-of-pocket costs are associated with increases in medication nonadherence.

And when seniors don't take their prescriptions as directed, their conditions worsen, and they often require more expensive medical care down the line. In fact, medication nonadherence is responsible for at least 10 percent of hospitalizations across the country. It also costs our healthcare system up to $289 billion every year.

Lawmakers want to improve Part D for these patients. In their draft bill, they've proposed eliminating beneficiary out-of-pocket costs for seniors who hit Part D's so-called "catastrophic phase"—which means they have spent at least $5,100 on prescription drugs for the year.

Capping patient out-of-pocket spending would provide some serious relief to these beneficiaries. Seniors will know they have a financial safety net should they fall ill and require one or more expensive medications. And they'll know they can afford to continue taking these medicines as directed, which improves overall health outcomes.

This new legislation is still in its infancy stages, so lawmakers will have to work out the nitty-gritty details of the reform in the coming weeks. But the draft discussion legislation shows a lot of promise—and it's one that patients and lawmakers alike should support.

New Policy Proposal Could Save Thousands of Lives per Year

Originally published July 2, 2019

Roughly 125,000 Americans will die this year as a result of not taking their medications. This phenomenon—known as "medication nonadherence"—is the cause of 10 percent of all hospitalizations nationwide. It also costs our healthcare system up to $289 billion annually.

The reason patients don't stick to their prescription regimen is simple: They can't afford to. Right now, nearly 25 percent of Americans taking prescription drugs struggle to afford their medications. And due to high out-of-pocket costs, 30 percent of patients had to forgo taking their medication at some point within the last year.

This is a problem—and one that is easily preventable. Lowering patient out-of-pocket spending on prescription drugs would improve medication adherence, boost patient health outcomes, and generate billions in savings.

Thankfully, there are a number of promising reforms in Washington. Some of those reforms target patient out-of-pocket spending in Medicare's Part D prescription drug benefit. This program currently helps more than 40 million seniors and people with disabilities afford their prescription drugs.

Right now, there is no cap on how much Part D beneficiaries could pay out-of-pocket for their prescriptions. This is a huge problem for beneficiaries battling one or more chronic diseases—many of whom require multiple drugs to live normal, healthy lives.

More than 800,000 Part D beneficiaries spend more than $5,000 per year on prescription medications—less than 2 percent of all enrollees. This 2 percent, however, accounts for 20 percent—roughly $3 billion—of enrollees' total out-of-pocket drug spending.

One proposal from the Trump administration would create an annual out-of-pocket maximum on prescription drugs. Once a beneficiary has paid $5,100 out of pocket at the pharmacy counter, he would no longer be on the hook for paying for his necessary medications for the rest of the year. For those patients who require costly, specialty drugs, those savings could add up quickly.

Another reform, also proposed by the Trump administration, would crack down on the greedy practices of pharmacy benefit managers, or PBMs.

PBMs work as middlemen between drug manufacturers and insurers. They negotiate large discounts on prescription drugs—sometimes 30 percent off a drug's initial list price.

Unfortunately, patients don't always see these savings. And as a result, industry profits go up while patient cost sharing remains the same.

President Trump's proposal puts a stop to this malpractice. Under his plan, it would be illegal for PBMs to keep any manufacturer rebates for themselves. These savings would instead flow directly to patients.

Patients would benefit greatly from this proposal. If the discounts negotiated by PBMs were re-directed to Part D beneficiaries at the point of sale, seniors would save up to $28 billion at the pharmacy over the next decade.

The federal government, meanwhile, could save up to $73 billion over the same time period.

The solution is simple: Americans are far more likely to take their medicines when they can afford them. And when patients follow their prescription regimen, they have better health outcomes as a result.

Capping out-of-pocket spending for Part D beneficiaries—and putting a stop to the predatory practices of PBMs—will save lives. Patients and lawmakers alike should fully support these reforms.

The White House's About-Face on Drug Rebates Is a Loss for Public Health

Originally published July 11, 2019

Even in the complicated ecosystem of drug pricing, one fact stands out: $166 billion in discounts from pharmaceutical companies go directly into the coffers of pharmacy benefit managers (PBMs). That's 37 percent of our nation's entire expense on drugs.

Not a single dollar of that largesse is used to reduce patients' out-of-pocket costs when they need medicines. So when the White House boldly developed a rule to change the dynamic by banning many rebates drug companies pay to PBMs under Medicare, policy experts applauded. That proposed rule died last night, the victim of intense lobbying and general ignorance. Who loses? Patients. Who wins? The status quo.

When a group of pharmaceutical CEOs testified before the Senate Finance Committee in February, Pfizer CEO Albert Bourla said he supported "reforms that would create a system in which transparent, upfront discounts benefit patients at the pharmacy counter, rather than a system driven by rebates that are swallowed up by companies in the supply chain." When asked if they would lower prices if the PBMs played fair, every hand on the panel went up.

Now the pharmaceutical industry is off the hook. So are the big payers. And important systemic change is off the table and the status quo rules. What happens to healthcare reform when all we're left with are silly soundbite solutions like "drugs from Canada" or "price controls from Slovakia"? That's no win for patients.

For shame, Mr. President. Was all of that really just political theater?

Pharmaceutical company rebates to PBMs that are tied to formulary restrictions create an incentive for entrenched market leaders to "bid" incremental rebates to prevent or limit access to competitive medicines. This model, coupled with escalating cost-sharing requirements, harms patients by driving up prices, which results in reducing access to innovative drugs.

Allowing PBMs to continue with business as usual means a continued disincentive to promote a more aggressive uptake of both biosimilars and less expensive generic drugs. Worse, reinforcing the status quo moves us even further away from a healthcare ecosystem based on competitive, predictable, free-market principles and not outrageous solutions such as "Medicare for All."

Zany ideas don't solve complicated public health problems. There are no simple solutions to complex obstacles—and politicians hate that.

Not following through on the proposed rule to ban rebates is harmful to patient health and the public purse. One of the biggest threats to the body politic is nonadherence to the medicines physicians have prescribed: It causes 125,000 deaths each year and is responsible for 10 percent of hospitalizations. Why don't people take their medicines? Often because their co-pays and co-insurance rates are too high.

Those rates aren't set by pharmaceutical companies. They're the domain of the PBMs and insurance companies. During the last five years, pharmaceutical spending has increased by 38 percent while the average individual health insurance premium has increased by 107 percent. During the same period, rebates, discounts, and fees paid by the biopharmaceutical industry to insurers and PBMs have risen from $74 billion to the aforementioned $166 billion. Facts, as John Adams said, "are pesky things."

Government policies should encourage rebate dollars to flow back to patients who need to take prescription drugs. Will greater transparency of contracting practices on the state level drive better PBM behavior? That's one theory. Such transparency efforts in New York and Connecticut, for example, will be the bellwether. But greed often trumps shame and, without penalties, will the C-suite at Big Payer choose to do the right thing by patients and reduce their hefty profits?

At the heart of the debate is whether we are going to improve our healthcare system through the use of smart and evolving free-market principles, such as more focused regulation that addresses the exclusionary contracting that locks out savings from biosimilars, or go down the sound-bite-laden path of "free healthcare."

All of this is contingent on the executive branch and Congress being honest brokers and not hucksters. As the great healthcare philosopher Frank Douglas once said to me, "It's not what you control, it's what you contribute."

Taking the heat off of PBMs does nothing to enhance access to essential medicines. The White House's decision may be a win for the status quo, but it is a lost opportunity for real systemic change.

CHAPTER SIX

Helping Patients

There are many issues (yes, even beyond drug pricing!) that comprise healthcare reform and deserve to be part of the public debate. But, dear reader, your time is too valuable and my knowledge too limited to cover everything in one book (or even an encyclopedia). There are, however, a few matters that are worthwhile to call to your attention and that you should call to the attention of anyone asking for your vote. After all, as Dr. Martin Luther King Jr. warned us, "Nothing in the world is more dangerous than sincere ignorance and conscientious stupidity."

Cannabis Happens

Originally published August 10, 2018

One of the positive side effects of the opioid crisis is a renewed (and long overdue) focus on new ways to advance pain management.

One therapy, cannabis, has long been discussed as a legitimate therapeutic alternative—but has been caught up in the debate over its nonmedical uses, along with its cousin, hemp-derived cannabidiol (CBD). Regardless of where you may stand on these issues, we are moving forward. The times they are a-changing. And, as with many advances in healthcare policy, states are presenting themselves as the laboratories of invention.

In New York, for example, cannabis is now a legal alternative to opioids. The policy, announced July 12, is a result of emergency regulations filed by New York State Health Commissioner Howard Zucker. Per Zucker, medical cannabis "has been shown to be an effective treatment for pain that may also reduce the chance of opioid dependence" and that offering providers treatment options other than opioids "is a critical step in combating the deadly opioid epidemic affecting people across the state."

The new policy allows patients who have been prescribed opioids to request cannabis as an alternative treatment and adds several new qualifying conditions. The broadened scope of conditions will now cover patients with severe pain that is not classified as "chronic," and also includes a provision allowing patients with opioid use disorder to use medical cannabis if they are enrolled in a certified treatment program. As part of the new program, patients and caregivers will receive ID cards that can be used along with a government-issued photo ID at registered dispensing facilities.

In addition to the new disease indications, the regulations are designed to increase certifying authority to nurse practitioners and physician assistants and allows for the approval of five additional organizations to manufacture and dispense. Home delivery will also be permitted.

While the emergency regulations are temporary, the New York State Department of Health has filed to adopt them permanently. Currently, cannabis is legally used by more than 62,000 patients via about 1,700 registered practitioners in the existing New York program. Health officials anticipate a reduction in opioid use and dependence, as well as a significant boost to state coffers, with legal cannabis sales expected to reach between $50 million and $70 million in 2018 (up from $20 million last year).

But it's important to understand that there's no benefit without some risk—and using cannabis for pain management and other disease mitigation isn't risk-free. Some key issues include:

No current standard in quality or production

There are few guidelines when it comes to how cannabis plants must be raised for dispensaries that sell the drug to patients. Each plant could be vastly different from another grower's plants, which in turn means that the buds will likely have very different levels of tetrahydrocannabinol (THC) or CBD. A joint rolled from one plant will provide a different intensity of a high than another plant, and there is no monitoring of the patient's use of the drug to ensure that they are finding the right type of plant for their needs.

No dosing standard

When patients are prescribed any other type of medication, they are given a dosing schedule by the doctor telling them how much to take, how to take it, and how often. When people are prescribed cannabis, they get a card that allows them to access dispensaries that sell the drug. They are not given any guidelines about how they should take it or in what amounts—something that would never happen with any other medication.

Potential for help and harm through chronic use

Chronic smoke inhalation and overdose on edibles are just two of the risks of chronic use of cannabis that we know about. Use of the drug legally for medicinal or recreational purposes has not been studied heavily, so we don't know the extent of the harm that can come to those who take the drug for long periods of time and/or in large amounts. The California, Colorado, and Nevada experience hasn't yet impacted rates of use/abuse. Watch this space.

Easier access = increased cannabis abuse

When it becomes easier for people to get cannabis, it means that they use more and have more in their homes. This in turn means that those who don't have medical cannabis cards—including teens and young adults—have increased access to the drug and may be more inclined to use and abuse it, developing a drug dependency when they may not have otherwise.

Legalization changes public opinion of harm potential

As more and more states legalize cannabis for medicinal—or recreational—purposes, it gives the impression to those who don't take the time to do the research that all doses of cannabinoids are harmless. In fact, for all drugs prescribed by a doctor, even prescription pills that are highly addictive such as OxyContin and Percodan, the prevailing attitude is that the doctor's OK makes them safe to use in any way or combination. Many don't realize that a doctor's prescription comes with guidelines for usage that must be followed for maximum safety and that since cannabis rarely even comes with this assistance, it is very important "proper use" is supported by education and technology.

More options are better—but as the saying goes, if you can't measure it, then it doesn't count. Broadening cannabis's legal bona fides for pain treatment isn't the end of the debate; it is only the beginning. Now we must develop ways to measure its effectiveness and develop ways to capture the real-world evidence that must drive evolving best medical practice and reimbursement policies.

In the immortal words of General George C. Marshall: "When a thing is done, it's done. Don't look back. Look forward to your next objective."

Stay tuned.

FDA and Juul E-Cig Regs Should Keep Antismoking Zealots at Bay

Originally published November 16, 2018

This week FDA Commissioner Scott Gottlieb announced a plan to close the on-ramp to kids using e-cigarettes while preserving adult use of "potentially less harmful forms of nicotine delivery." The plan's centerpiece curbs the sale of flavored e-cigarettes, which regulators claim entice youth (logic that apparently doesn't extend to Absolut Peach vodka or Jack Daniel's cinnamon whiskey), but which also help smokers quit because they don't remind them of traditional cigarettes.

The problem is that antismoking zealots want to go further and remove e-cigarettes from the market altogether while leaving most tobacco products untouched. That would snuff out the most impactful public health development in decades. The FDA's action is designed to reduce e-cigarette use among high school and middle school students "because there is also evidence that a large percentage of these children will become addicted to nicotine and ultimately take up smoking." Yet plummeting and all-time low smoking rates—especially among young adults—would suggest otherwise. In contrast, the FDA recognizes that use of e-cigarettes can help adults stop smoking.

Days before the FDA's announcement, e-cigarette manufacturer Juul, which commands about 75 percent of the e-cigarette market share, self-imposed regulations to curb teen use of its products. Juul's "Youth Prevention Plan" will eliminate pod flavors, such as cucumber and mango, from the 90,000 or so retail outlets across the country.

Flavored pods will still be available online, but Juul is ratcheting up its age verification tools to ensure that purchasers are 21 years of age or older. These include demanding permanent addresses, the last four digits of social security numbers, pictures of government-issued ID, and real-time photos.

To prevent bulk purchases that feed black markets that target minors, Juul introduced new purchase limits and product serialization. It is also quadrupling its secret shopper program to ferret out rogue retailers.

And finally, for good measure, Juul is spiking its social media pages, which teens are more likely to visit.

To be sure, faced with an existential threat of lawsuits and an outright ban from regulators and anti-e-cigarette zealots on a warpath, the e-cigarette industry needed to respond. Yet in doing so it means e-cigarette manufacturers will save fewer lives. Juul's actions, lauded by the FDA, will improve the public risk vs. benefit tradeoff of e-cigarettes in the short term by possibly increasing tobacco use in the future. Gottlieb has acknowledged that the speed bumps Juul has put in place will make it harder for consumers over the age of 21 to buy e-cigarettes. For instance, those reliant on flavored e-cigarettes to quit will be out of luck next time they go to the store for their nicotine fix. Yet if these consumers understandably don't want to go through the rigmarole and privacy invasion of online purchases, they may simply pick up a pack of cigarettes instead.

Similarly, while some limit on bulk purchases is probably smart, Juul's new limit of 10 per year seems low for former smokers who want to gift e-cigarettes to friends and family trying to quit smoking.

E-cigarettes have helped millions of Americans quit smoking. And they have helped millions more reduce their consumption. According to one independent study, 70 percent of U.S. users who had been smokers reported quitting following their e-cigarette use. And an incredible 96 percent of the former smoking e-cigarette users reported being less interested in smoking. According to randomized trials, quit rates are directly correlated with e-cigarette use. And the frequency of e-cigarette use increases quit success.

With smoking still responsible for one in every five deaths, Boston University preventive-medicine physician Michael Siegel believes e-cigarettes are the singular technology that could put an end to smoking. Yet the anti-e-cigarette zealots want to ban e-cigarette sales altogether.

Ironically, it's young adults who have the most to lose. Centers for Disease Control data show that the 18- to 24-year-old demographic has seen by far the largest decline in smoking in recent years. This is partially thanks to increased e-cigarette use among kids under 18. Barring people from buying e-cigarettes for their first three years of adulthood (18–20 years old) will likely increase the number of cigarette consumers while reducing the number who are able to quit.

Public health advocates concerned about rising use of e-cigarettes among those under the age of 18 should recognize that e-cigarette use is often a symptom of a deeper emotional turmoil such as anxiety or depression, which are skyrocketing among youth. While self-medication with e-cigarettes isn't ideal, it is better than doing so with marijuana, alcohol, or traditional cigarettes.

Juul's actions and the FDA initiative are a historic attempt to harness the public health benefits of e-cigarettes by maximizing the number of people who use them to stop smoking and limiting the number of people who use them to take up tobacco. This laudable effort is not perfect and, as we note, must be monitored to ensure it doesn't make smoking more likely.

Yet judging by their dismissive response to Juul's actions, antismoking zealots don't seem to really care about those dying from cigarettes. If they did, they would champion e-cigarettes for the public health phenomenon that they are. Perhaps they're worried that if Big Tobacco goes out of business because smokers turn to e-cigarettes instead, so will they.

Health Sharing, a Viable Option for Health Care

Originally published January 8, 2019

When it comes to the intense (and intensifying) debate over health care in America, the conversation begins at the top of the alphabet. "A" is for "Access."

One of the key lessons we've learned from the Obamacare experience is that "having health insurance" isn't the same thing as having access to quality health care.

Millions of Americans are learning the hard way about the tradeoffs between health insurance policies that have no premiums (so-called "free health care") or those with low premiums but high out-of-pocket costs. The dual promises of "free health care" and the ability to "keep your existing insurance if you want to" were quickly proven to be mere political slogans with no basis in reality.

"The truth," as Oscar Wilde quipped, "is rarely pure and never simple." Nowhere is this truer than when it comes to insurance design. Americans are understandably "B" (baffled) about their lack of "C" (choice).

Broader access comes via expanded choice. Some believe average citizens can't be allowed to choose their health coverage. These are the same voices who initially decried the Medicare Part D drug benefit because seniors couldn't possibly choose the plan that was best for them. Today Part D has a 90 percent approval rating among those 65 and older.

In August 2018, the Trump administration finalized rules that expand access to so-called "skinny" short-term health insurance plans. These are low-cost, low-coverage options designed for healthy young people — among others — who chose to opt out of Obamacare. They instead prefer to pay the mandated fine, which is generally lower than the price of even a low-cost Obamacare "bronze" level policy.

Another option that is gaining momentum among wide demographic swaths is health-sharing ministries — organizations that facilitate the sharing of health care costs among individual members who have common ethical or religious beliefs.

A health care sharing ministry does not use actuaries, does not accept risk or make guarantees, and does not purchase reinsurance policies on behalf of its members. The monthly cost of membership in a health care sharing ministry is generally lower than the cost of insurance rates for a similar policyholder or family.

According to an estimate from the Alliance for Health Care Sharing Ministries, when the Affordable Care Act — which recognized health care sharing ministries as an alternative form of insurance — fully went into effect in 2014, an estimated 160,000 people were enrolled in these programs nationwide. Today, as many as 1 million people have joined, says the alliance.

Health Sharing eliminates traditional insurance marketing expenses. The Affordable Care Act mandated that health insurance companies use a maximum of 40 percent of health insurance premiums for marketing and administrative costs. Health sharing plans are, by definition, benevolent organizations structured for the benefit of their members, with administrative costs generally in the 10 percent range.

Members of a health share pay a predetermined monthly amount, which goes to pay other members' eligible medical costs. At its core, health sharing is simple and encompasses the basic belief of caring for one another, as well as the time-honored ideals of strength in numbers and sharing burdens together.

Thirty states have safe harbor laws that distinguish health care ministries from health insurance organizations. Some of the larger health care sharing ministries include Christian Healthcare Ministries, Medi-Share, a program of Christian Care Ministry, Samaritan Ministries, Liberty HealthShare, MCS Medical Cost Sharing and Altrua HealthShare.

Most health sharing ministries tend to have restrictions, such as abstaining from excessive drinking and use of tobacco or illegal drugs. They usually require members to be in good health and make a statement of belief, as well. Liberty HealthShare is more inclusive, accepting members with a wide variety of beliefs.

According to Larry Foster, CEO at Liberty HealthShare, "Health Sharing is an ingenious answer to the high cost of today's healthcare."

"Ingenious" is in the eye of the beholder, but health sharing is an option that needs to be more widely understood and considered by millions of Americans seeking alternative access to health coverage. More options are better and Americans (yes — even religious ones) have the smarts to make their own choices.

"Liberty is the right to choose. Freedom is the result of the right choice," said Jules Renard.

Is "Right to Try" Really "Right to Pay?"

Originally published January 31, 2019

More than 1.7 million Americans will be diagnosed with cancer this year. Almost one in three of them will eventually succumb to it. Knowing these odds, patients are eager to try anything to improve on them. People with cancer and their loved ones want quicker access to the latest medicines— even if they are still at an experimental stage.

That's why so many Americans applauded passage of the Right To Try Law, which President Trump signed in May. The measure creates a legal pathway for people with terminal illness to access new medications that have only been through one phase of Food and Drug Administration (FDA) approval.

While well intentioned, the law could create opportunities for hucksters and snake-oil salesmen to prey on those who are vulnerable. As the FDA develops its protocols for right to try, the agency must ensure that patients are protected from anyone trying to game the system.

Consider the case of BrainStorm Cell Therapeutics, a drug company with a new experimental stem cell treatment called NurOwn. The treatment is for amyotrophic lateral sclerosis, the terrifying disease that killed baseball superstar Lou Gehrig. Those suffering from the neurodegenerative disorder gradually but inexorably lose all ability to control muscle movements. Eventually, patients are completely isolated in their immobile bodies—typically until their breathing stops.

NurOwn is harvested from the stem cells of each individual patient. It's currently under development, and results to date have been inconclusive. A midstage study involving 48 participants found those given the drug did appear to respond, though for most the benefit didn't last.

It's highly uncertain whether NurOwn will eventually pass the FDA's scientifically rigorous safety and efficacy standards. Nevertheless, NurOwn's developer sought to make it available to patients. But here's the catch: "Right to try" doesn't mean "right to try for free." And that's where BrainStorm Cell Therapeutics had its own brainstorm: Proposing to sell its experimental treatment for hundreds of thousands of dollars.

Insurance doesn't generally cover treatments that have not been approved by regulators or proven to work in clinical trials. That means patients would have to pay for the therapies out of their own pocket. Some could be desperate enough to sell their house to come up with the money.

Right-to-try legislation isn't designed for commercial profit. Yet after Trump signed the bill, requests for NurOwn skyrocketed and BrainStorm's shares closed up 2.8 percent.

BrainStorm ultimately bailed out on offering NurOwn under the right-to-try law, but only after a maelstrom of controversy. Yet the company's initial impulse was likely only the first of many potentially unsafe and exploitative actions, which the FDA must act to prevent.

No one actually knows if NurOwn works. But desperately ill patients may gamble everything to try it anyway. Such are the unintended consequences of right-to-try.

No wonder the American Society of Clinical Oncology (ASCO) opposes right-to-try legislation on both the state and federal levels.

"ASCO supports access to investigational drugs outside of clinical trials, when adequate patient protections are in place," ASCO chief medical officer Richard Schilsky, MD, said. "We don't support right-to-try legislation, however, because these laws ignore key patient protections without actually improving patient access to investigational drugs outside of clinical trials."

Let's hope BrainStorm—and other likeminded companies—have learned that cashing in on the uncertain hopes of desperate patients is an unacceptable industry practice. And let's make sure the FDA sends that message in no uncertain terms.

Understanding the FDA's Temp on Hemp

Originally published March 18, 2019

If you're living with cancer, you're probably no stranger to pain or, most likely, to opioids.

One therapy, cannabis, has long been discussed as a legitimate therapeutic alternative, but it has been caught up in the debate over its nonmedical "recreational" use, along with its cousin, hemp-derived cannabidiol (CBD). Regardless of where you may stand on these issues, we are moving forward. The times they are a-changin'.

Yet the reality is more complicated than the sound bites, and the reality is that it's very complicated. Nowhere is this more obvious or urgent than when it comes to the role and responsibilities of the Food and Drug Administration (FDA) and its regulation of CBD.

Last December, the Agriculture Improvement Act of 2018 ("the Farm Bill") was signed into law. Among other things, this new law changed the jurisdiction of certain federal authorities relating to the production and marketing of hemp. These changes include removing hemp from the Controlled Substances Act, which means that it will no longer be an illegal substance under federal law. But just as important for the FDA is what the law didn't change: Congress explicitly preserved the agency's current authority to regulate products containing cannabis or cannabis-derived compounds, such as CBD.

Specifically, it remains unlawful under FDA regulations to market CBD products as dietary supplements, regardless of whether the substances are hemp-derived. Per the FDA, this is because CBD is an active ingredient in an FDA-approved drug (Epidiolex) that has been the subject of substantial clinical investigation. Under the Federal Food, Drug and Cosmetics Act, it's illegal to introduce drug ingredients like these into dietary supplements. So Farm Bill or no Farm Bill, the FDA has made it clear that hemp-derived cannabidiol is not a legal dietary supplement.

According to the FDA, "We'll take enforcement action needed to protect public health against companies illegally selling cannabis and cannabis-derived products that can put consumers at risk and are being marketed in violation of the FDA's authorities."

In February 2019, FDA Commissioner Scott Gottlieb, MD, announced a newly created Botanical Safety Consortium, a public-private partnership that will gather leading scientific minds from industry, academia, and government to promote scientific advances in evaluating the safety of botanical ingredients and mixtures in dietary supplements. This group will look at novel ways to use cutting-edge toxicology tools, including alternatives to animal testing, to promote the goals of safety and effectiveness the agency shares with consumers and other stakeholders. No mention was made of hemp-derived CBD—but that's certainly viewed by many as the 800-pound gorilla sitting in the regulatory corner.

According to the FDA announcement, "Our first priority for dietary supplements is ensuring safety. Above all else, the FDA's duty is to protect consumers from harmful products. Our second priority is maintaining product integrity: We want to ensure that dietary supplements contain the ingredients that they're labeled to contain, and nothing else, and that those products are consistently manufactured according to quality standards. Our third priority is informed decision-making. We want to foster an environment where consumers and health care professionals are able to make informed decisions before recommending, purchasing or using dietary supplements."

What's the relevance for the nascent but swiftly growing CBD industry? First, that its marketing plans need to be put on pause now that the FDA has stepped up to the plate. The agency carries a big stick and knows how to use it. Next, as with all FDA-regulated products, manufacturing quality and labeling integrity are joined at the hip, inseparable *condiciones sine quibus non*. Complying with FDA regulations isn't easy or inexpensive and is certainly not a mom-and-pop proposition.

Many people with cancer have enjoyed the benefits of CBD oils. One study showed that CBD extract is effective for patients with advanced cancer pain not fully relieved by strong opioids. But long-term, broad-based, and validated patient outcome studies don't exist—at least not yet.

And the FDA isn't practicing salutary neglect. Many in the cannabis community think this is regulatory creep on the part of the FDA, and maybe it is. But that doesn't change the law or the facts on the ground. Waving away as "Big Brotherism" the important public health role of the FDA doesn't make the agency's position or authority any less real or relevant.

It's time for the proponents of CBD—including many highly vocal patients, physicians, pharmacists, manufacturers, and distributors—to become part of the solution and participate in the FDA's new Botanical Safety Consortium. Remember what they say in Washington, "If you're not at the table, you're on the menu."

Contaminated Generic Drugs Reveal an Urgent Public Health Crisis

Originally published April 4, 2019

Quality can be stipulated, but it cannot be assumed. Quality occurs by design; it cannot be tested into a product. Nowhere is this truer or more urgent than when it comes to generic drugs. Why generics? Because they represent 90 percent of the volume of medicine consumed by Americans.

Medications approved by the U.S. Food and Drug Administration (FDA) rest upon three pillars: safety, efficacy, and quality. Up until recently, quality has been a silent partner—but that's no longer acceptable. It often takes a crisis to propel important but overlooked issues to the forefront. Witness the recent public health imbroglio over valsartan, a widely prescribed angiotensin II receptor blocker (ARB), most commonly used to treat high blood pressure. Over the past few months, a dozen companies have recalled their generic versions of valsartan and other ARBs because of contaminated ingredients containing three different potential carcinogens.

How can such things happen? Isn't a generic drug "identical" to its innovator reference product? Aren't generic drugs approved by the FDA because they are just as safe and effective as the original brand? Alas, the answers to these questions are complicated.

First, as to generic drugs being "the same as" or "identical to" their brand cousins, the answer is they are not. Generic drugs are pharmaceutically equivalent (same active ingredient, dose, and dosage form) and "bioequivalent" to the innovator brand. The fundamental regulatory assumption (as codified in the Hatch-Waxman Act) is that if two formulations are shown to be bioequivalent (resulting in equivalent blood levels of the drug), it is presumed that they will reach the same therapeutic effect, are therapeutically equivalent, and can be used interchangeably. Bioequivalent doesn't mean identical, and such differences are more important for certain types of medicine, specifically for medications that require a "narrow therapeutic index" for conditions such as epilepsy in which even a small difference can result in therapeutic failure.

Bioequivalence, however, was not the issue with valsartan. The problem was caused by contamination of the active pharmaceutical ingredients (API) manufactured in China and India. The FDA believes that genotoxic impurities may be generated by changes made in the API manufacturing process, but also may result from the reuse of materials, such as solvents.

According to the FDA's website, more than 282 warning letters have been sent to pharmaceutical companies related to manufacturing problems since 2015. Of those, 70 involved active pharmaceutical ingredients.

Generic drugs are too important to America's public health to allow anything less than complete confidence in their safety, efficacy, and quality. Manufacturers unaffected by contamination issues have increased the price of valsartan at least twofold amid the recent recalls; Alembic Pharmaceuticals more than tripled the price of 17 of its valsartan formulations, with price hikes ranging from 329 to 469 percent. The bad actors who take advantage of recalls must be put on notice that such behavior, while not illegal, is certainly not in the public's best interest.

This is why the FDA recently announced a plan to revamp its rules that govern how medicine is manufactured in an effort to ensure the safety of the nation's drug supply, as recalls of contaminated imports from developing countries widen. Per outgoing FDA Commissioner Scott Gottlieb, "We've seen a lot of instances of adulterated products—contamination, impurities—recently ... The underlying causes have been traced back to manufacturing, inadequate quality controls and generally poor management oversight." The proposed rule changes will focus on the active pharmaceutical ingredients; however, Gottlieb said the overhaul is only now in its early stages and may take several years to complete.

An important inside-baseball but mostly overlooked move by the FDA was to develop a specific pharmaceutical inspectorate within the agency's Office of Regulatory Affairs (ORA) rather than having district-based offices drive the process without the necessary authorities or resources.

According to Dr. Richard Kovacs, president of the American College of Cardiology, the recent ARB recalls have been "a bit of a reality check" in an era of easing regulatory constraints.

"If there is a silver lining to this, it's going to be a measured approach to how we protect our drug supply in a global economy," he said. "Certainly making everything in the U.S.—up to and including fill-and-finish—could be one solution, but that will have to be balanced against the cost of our drugs."

"Inexpensive drugs are going to mean inexpensive manufacturing processes, and you would hope there wouldn't be corners cut or quality controls that aren't going to be kept, but the reality is it's going to happen," Kovacs said. "I think this is a societal question and an important regulatory question as to what the right balance of safety and efficacy is in this new framework."

The bottom line is that generic manufacturers are going to be scrutinized more robustly and more regularly. Going forward, physicians and patients may start demanding products manufactured by specific generic companies that prioritize quality over cost.

As the *Journal of Infection* so aptly stated: "Nothing is more expensive than treatment failure."

How "Right to Try" Laws Fail Patients Who Need Experimental Medicine

Originally published April 24, 2019

It's time to explore the evolution of the "Right to Try" movement. Originally presented as a way for desperately ill patients to gain access to developmental medicines outside of the Food and Drug Administration (FDA) process, Right to Try laws have proven to be a dud. After all the fuss and bother by politicians and pundits, the number of patients receiving treatment via this new pathway can be counted on the fingers of one hand—with enough fingers left to type this commentary.

To sum it up, the Right to Try movement is failing because it is based on four fallacies.

Fallacy No. 1: The FDA Was a Bottleneck to Patients Seeking Experimental Treatment

Many have blamed the FDA for impeding patients' searches for experimental treatment options, but the facts speak otherwise. Over the last five years, the FDA has authorized more than 9,000 applications across drugs, biologics, and devices through the agency's Expanded Access program. Furthermore, the agency has authorized approximately 99 percent of all the requests received, across all application types.

The fact that the FDA's volume of Expanded Access applications has increased since the passage of state and federal Right to Try legislation speaks to the robustness of the agency's program and the importance of the Right to Try concept. However, it isn't a tremendous leap of faith to say that it's been mainly media attention that has helped communicate to patients, caregivers, and physicians this better, speedier FDA pathway.

Fallacy No. 2: Right to Try Means Drug Developers Are Required to Provide Their Medicines to Patients Upon Request

According to Senator Ron Johnson (R-Wis.), one of the prime forces behind the federal Right to Try Act, "Patients with terminal diseases ought to have a right to access treatments that have demonstrated a level of safety and could potentially save their lives." Fine words, but this isn't what the actual legislation calls for; drug developers can accept or deny a request under Right to Try laws for any reason. "Right to Ask" doesn't have the same rhetorical resonance, but that's the reality.

Right to Try also doesn't mean patients have the right to try—for free. While, the FDA limits what can be charged for an experimental medicine (raw materials, labor, and nonreusable supplies and equipment needed to make the quantity of drug required for the patient's use—or costs to acquire the drug from another manufacturing source, plus shipping and handling), Right to Try allows pricing to be a potential profit center.

Fallacy No. 3: Right to Try Provides Hope to Desperately Sick Patients

Claiming to facilitate access isn't hope. It's false hope. It's cruel. In January 2008, the U.S. Supreme Court, without comment, opted not to accept an appeal of Abigail Alliance for Better Access to Developmental Drugs v. von Eschenbach. In other words, the federal appeals court ruling that patients do not have a constitutional right to experimental drugs stands.

Not clarifying that Right to Try laws aren't a "guarantee" of access is, at best, an unfortunate oversight. At worst, it's a lie perpetrated by FDA-bashers to further their broader counter-regulatory agenda. For shame. False hope won't help convince payers (both public and private) to reimburse for experimental treatments acquired through Right to Try. There is no rationale for a "hope placebo," and divine intervention falls outside of legislative or regulatory authorities and is not likely to result in a more compassionate reimbursement philosophy.

Fallacy No. 4: Access to Experimental Medicine Is Only About an Individual Patient

By far the most daunting of the four fallacies, Right to Try laws aren't specifically concerned with an individual patient's unique situation. Couched in the language of personal rights and liberties, the Right to Try movement, and such legislation, never once mentions the value of the evidence derived from patients using experimental medicines outside of their FDA-vetted registration trials. This is crucial. Whether or not any one patient achieves a successful therapeutic outcome via such access, it's critical to capture the data from all patients. Knowledge is power, and nowhere is this truer than in how desperately ill patients respond to experimental medicines. Data counts. To paraphrase the management guru Peter Drucker, the information revolution will shift from the generation of data to figuring out the meaning and purpose of the data with the patient's perspective in mind.

As Patty Delaney, the former director of the FDA's cancer liaison program, explained in 2007, "The patient has a right to be heard, but in the end, it's the data that matters. FDA opinions about safety and efficacy are always based on data."

The debate over access to experimental medicines must continue to evolve from Right to Try to the right way to try. Hope must be facilitated. Science must be the pathway. Politics and posturing must be ignored. Reality must be acknowledged. As science fiction author Philip Dick reminds us, "Reality is that which, when you stop believing in it, doesn't go away."

Sunshine and CBD

Originally published May 14, 2019

On May 31, the Food and Drug Administration (FDA) will hold a meeting to discuss "Scientific Data and Information About Products Containing Cannabis or Cannabis-Derived Compounds." Specifically, this means CBD (cannabidiol). In short, the FDA has realized that it's time to face reality.

Reality is more complicated than sound bites. The absence—to date—of advanced regulatory thinking relative to CBD has resulted in a maelstrom of false claims and shoddy quality standards. Nature abhors a vacuum.

The FDA has made it clear that hemp-derived CBD is not a legal dietary supplement.

According to the FDA, "Our first priority for dietary supplements is ensuring safety. Above all else, the FDA's duty is to protect consumers from harmful products. Our second priority is maintaining product integrity: We want to ensure that dietary supplements contain the ingredients that they're labeled to contain, and nothing else, and that those products are consistently manufactured according to quality standards. Our third priority is informed decision-making. We want to foster an environment where consumers and health care professionals are able to make informed decisions before recommending, purchasing or using dietary supplements."

Bravo, but what's the relevant messages for the nascent but swiftly growing CBD industry and the American consumer? First, that aggressive and misleading marketing campaigns need to be put on pause now that the FDA has stepped up to the plate. Next, as with all FDA-regulated products, manufacturing quality and labeling integrity are joined at the hip, inseparable *condiciones sine quibus non*. Complying with FDA regulations isn't easy or inexpensive and is certainly not a mom-and-pop proposition.

Many in the CBD community think this issue is one of regulatory creep on the part of the FDA, but that doesn't change the law or the facts on the ground. Waving away as "Big Brotherism" the important public health role of the FDA doesn't make the agency's position or authority any less real or relevant.

It's time for the proponents of CBD—including many highly vocal patients, physicians, pharmacists, manufacturers, and distributors—to become part of the solution. If you're not at the table, you're on the menu.

Some key issues include:

No current standard in quality of production

We mustn't repeat the tragic flaws of limiting the FDA's hand via outdated DSHEA (the legislation that requires dietary supplements to be regulated as foods). Quality must always trump corporate convenience.

No dosing standard

When patients are prescribed any other type of medication, they are given a dosing schedule by the doctor telling them how much to take, how to take it, and how often. When people are told to use CBD by physicians, pharmacists, or friends, they are not given any peer-reviewed guidelines about how they should take it or in what amounts—something that would never happen with any other medication.

Potential for help and harm through chronic use

What does serious research tell us? Hardly anything, and the plural of *anecdote* isn't *data*. We mustn't repeat the mistakes that led to the opioid epidemic.

Legalization changes public opinion of harm potential

If you can't measure it, then it doesn't count. Quantifying CBD's therapeutic and manufacturing bona fides for pain treatment isn't the end of the debate; it is only the beginning. Now we must develop ways to measure its effectiveness and develop ways to capture the real-world evidence that must drive evolving best medical practice and reimbursement policies.

It's time for the FDA to assume the mantle of leadership and work with the CBD community to develop solid standards and practices for manufacturing and labeling quality, sound science, and practical consumer education. As President Kennedy said, "Leadership and learning are indispensable to each other."

CONCLUSION
Next Steps

"Let us overthrow the totems, break the taboos. Or better, let us consider them cancelled. Coldly, let us be intelligent." – Pierre Trudeau

The debate over drug pricing has devolved into the repetition of the same old shibboleths and political talking points. In the words of James Thurber, "There are two kinds of light—the glow that illuminates, and the glare that obscures."

We must all embrace workable and forward-looking solutions such as increasing the use of less costly generics and biosimilars, increasing competition in the brand marketplace, calculating spending based on net prices, eliminating payer demands that patients pay more for their medicines through their insurance rather than if they paid in cash, and, most urgently, demanding greater ecosystem transparency so that patients can benefit directly from the slew of rebates, concessions, and discounts offered to insurance companies and PBMs.

Shortly before his death, I had the privilege of a private meeting with Nobel Laureate Joshua Lederberg. We talked about the state of American healthcare and a host of future-oriented issues. At the end of the meeting he put everything into perspective in a single sentence. He leaned over the table and said, "The real question should be, is innovation feasible?"

Innovation equals hope. When a politician asks for your vote ask them how their healthcare policies impact the future of pharmaceutical innovation and patient-centric care. If they stumble, send them a copy of this book—and I'll pay for it.

Peter J. Pitts
June 24, 2019
New York City

"In this work, when it shall be found that much is omitted, let it not be forgotten that much likewise is performed." –Samuel Johnson, *A Dictionary of the English Language*

BIBLIOGRAPHY

CHAPTER ONE: DRUG DEVELOPMENT IN THE UNITED STATES

- Debunking the Myth of Me-Tooism
  Originally published in The Orange County Register, *November 14, 2006*
- Four Bucks: Cheez Whiz and Imitation Prozac
  Originally published in The Baltimore Sun, *November 22, 2006*
- Alzheimer's Setback Shows Difficulty of Drug Development
  Originally published in Reuters, *September 23, 2010*
- The High Cost of Slashing Patents
  Originally published in The Buffalo News, *May 6, 2013*
- Federal Funds Should Go to Medicine-Makers
  Originally published in The Chicago Tribune, *May 11, 2014*
- Disarming the Patent Death Squad
  Originally published in The Wall Street Journal, *June 11, 2015*
- 360° Trade
  Originally published in Morning Consult, *June 13, 2016*
- A Flawed Study Depicts Drug Companies as Profiteers
  Originally published in The Wall Street Journal, *October 9, 2017*
- Elizabeth Warren's Threat to Medical Progress
  Originally published in Real Clear Health, *January 6, 2019*
- Greedy Tort Bar Tarts Up the CREATES Act
  Originally published in The Hill, *February 12, 2019*
- Cornyn-Blumenthal Drug Bill Carries Serious Side Effects, No Benefits
  Originally Published in The Washington Times, *May 24, 2019*

CHAPTER TWO: DRUG IMPORTATION

- Is Hezbollah Filling Your Prescription?
  Originally published in The San Diego Union-Tribune, *October 13, 2006*
- A Poison Pill, Indeed
  Originally published in The New York Sun, *May 24, 2007*
- Paying the Ultimate Price for Cheap Pills
  Originally published in Prince George's Sentinel, *October 25, 2007*
- The GAO's Alarming Message
  Originally published in The Press-Enterprise, *February 25, 2008*

CHAPTER THREE: PART D REFORM

ABOUT THE AUTHOR

Peter Pitts is President of the Center for Medicine in the Public Interest. A former member of the United States Senior Executive Service, Pitts was FDA's Associate Commissioner for External Relations, serving as senior communications and policy adviser to the Commissioner.

His comments and commentaries on health care policy issues regularly appear in *The New York Times*, *The Chicago Tribune*, *The Washington Post*, and *The Wall Street Journal*, among others.

His book, *Become Strategic or Die*, is widely recognized as a cutting edge study of how leadership, in order to be successful over the long term, must be combined with strategic vision and ethical practice. He is the editor of *Coincidence or Crisis*, a discussion of global prescription medicine counterfeiting and *Physician Disempowerment: A Transatlantic Malaise*.

He is a Visiting Professor at the University of Paris, Descartes Medical School, a Visiting Lecturer at the École Supérieure des Sciences Économiques et Commerciales (Paris and Singapore), and has served as an adjunct professor at Indiana University's School of Public and Environmental Affairs and Butler University. A graduate of McGill University, he is married to Jane Mogel, and has two sons.